Department of Transport

ASSESSING THE ENVIRONMENTAL IMPACT OF ROAD SCHEMES

The Standing Advisory Committee on Trunk Road Assessment

Chairman: Mr D A Wood QC

March 1992

London Her Majesty's Stationery Office

London: HMSO

© *Crown copyright 1992*
Applications for reproduction should be made to HMSO
First published 1992

ISBN 0 11 551103 2

THE DEPARTMENT OF TRANSPORT

THE STANDING ADVISORY COMMITTEE
ON TRUNK ROAD ASSESSMENT
2 MARSHAM STREET LONDON SW1P 3EB

CHAIRMAN: MR DEREK WOOD QC

Rt Hon Malcolm Rifkind QC MP
Secretary of State for Transport

My Ref:

Your Ref:

1 November 1991

Sir

In September 1989 this Committee was invited to review the Department's methods for assessing environmental costs and benefits.

We now have pleasure in submitting our Report.

In carrying out our review we have received a large volume of evidence from the bodies and individuals referred to in Chapter 1 of and Annex 1 to our Report. We would like to acknowledge the willing help which we have received from your Department in providing us with the material and technical data which we have asked it to provide, and in making officials freely available to us to answer questions and give explanations of current procedures and practices. In addition we have had the benefit of able and efficient Secretarial assistance from our Technical and Administrative Secretaries, Mr S Grayson and Mr R Lamsdale. We wish to record our appreciation of the particular help which they have given us during the course of our work.

Yours faithfully,

D A Wood QC - Chairman

R H Stewart - Vice Chairman

H J Wootton

P B Goodwin

P J Hills

A M Lees

P J Mackie

CONTENTS

		Page Number
PART I:	**INTRODUCTION**	1
CHAPTER 1:	BACKGROUND TO THE REPORT	3
CHAPTER 2:	ROADS AND THE ENVIRONMENT	7
PART II:	**THE PLANNING OF TRUNK ROADS**	17
CHAPTER 3:	THE PLANNING OF TRUNK ROADS	19
	Identification Studies	19
	Entry into the Roads Programme	20
	Procedure after Entry into the Roads Programme	20
	Preferred Routes or Schemes and Public Inquiry	20
PART III:	**THE DEVELOPMENT OF CURRENT ASSESSMENT METHODS**	23
CHAPTER 4:	CURRENT APPRAISAL METHODS: TAM AND COBA	25
	Traffic Appraisal and Traffic Forecasting	25
	The COBA Program	25
	Valuation of Travelling Time	25
	Valuation of Accidents	26
	Vehicle Operating Costs	26
	Construction and Preparation Costs	27
	Cost of Maintenance	27
	Discounting	27
CHAPTER 5:	ENVIRONMENTAL APPRAISAL: THE ACTRA AND THE FIRST SACTRA REPORTS	29
CHAPTER 6:	CURRENT APPRAISAL METHODS: MEA, FRAMEWORKS AND COLOUR CODING	33
	The MEA	33
	The Framework	33
	The Environmental Appraisal	34
	Colour Coding	35
CHAPTER 7:	URBAN ROAD APPRAISAL: THE SACTRA 1986 REPORT	37

	Page Number
CHAPTER 8: ENVIRONMENTAL ASSESSMENT UNDER EC DIRECTIVE 85/337	41

PART IV: DISCUSSION OF CURRENT PRACTICE — **45**

CHAPTER 9: WHAT IS GOOD PRACTICE?	47
CHAPTER 10: OBJECTIVES-LED ENVIRONMENTAL ASSESSMENT	51
Setting Objectives and Targets	52
CHAPTER 11: THE PRESENTATION OF ENVIRONMENTAL STATEMENTS AND ENVIRONMENTAL ASSESSMENTS	55
The Timing of Assessment	55
The Statutory Requirements for Environmental Statements	56
Existing Environmental Statements	56
The Use of the Framework in Environmental Statements	58
Development of the Environmental Statement	58
Frameworks and Assessment Summary Reports	59
CHAPTER 12: REVIEW OF THE MANUAL OF ENVIRONMENTAL APPRAISAL	61
Additions to the Checklist of Impacts	62
Preservation and Mitigation	67

PART V: THE MONETARY VALUATION OF ENVIRONMENTAL EFFECTS — **69**

CHAPTER 13: VALUING THE ENVIRONMENT: PRINCIPLES	71
CHAPTER 14: REVIEW OF EXISTING VALUATION TECHNIQUES	75
Environmental Effects Which Cannot Sensibly be Valued in Money Terms	75
Actual Costs	75
Special Case (1): The Cost of Land-Take	76
Special Case (2): Mitigation which is not Carried Out	78
Values Revealed by Behaviour	78
Individual Values Expressed by Statements of Preference	80
Equity and Distribution	82
Values which derive from Policy Constraints and Targets	83
CHAPTER 15: THE WAY FORWARD	85
Actual Costs	85
Shadow Costs	85

	Page Number
Revealed Values, Contingent Valuations and Stated Preferences	86
Traffic Forecasting and COBA	87
Postscript	88

PART VI: CONCLUSIONS AND RECOMMENDATIONS — 89

CHAPTER 16: CONCLUSIONS AND RECOMMENDATIONS — 91

Introductory	91
Good Practice	91
The Timing of Assessment	93
Content of the Assessment	94
Guidance on the Assessment	94
Monetary Valuation of Environmental Effects	95
Traffic Forecasting and COBA	97
Presentation of the Assessment	97

ANNEX 1:	Evidence Received	99
ANNEX 2:	Terms of Reference and Membership	105
ANNEX 3:	The Main Stages in the Planning and Construction of a Trunk Road	109
ANNEX 4:	Manual of Environmental Appraisal: Part A Section 2 Appendix 1 Public Consultation Framework	117
ANNEX 5:	Manual of Environmental Appraisal: Part A Section 2 Appendix 2 Public Inquiry Framework	125
ANNEX 6:	Rules for the Calculation of Colour Coding	135

GLOSSARY — 141

List of Figures

		Page Number
Figure 1:	Increase in car ownership	8
Figure 2:	Examples of the wide range of environmental effects of trunk road development	11
Figure 3:	Environmental and social impacts: Summary of changes proposed for the assessment of urban main road schemes	39
Figure 4:	Suggested contents list for the Manual of Environmental Assessment	63

Cover photographs include: M26, M20 Intersection, M6 - Interfoto Picture Library Ltd.

PART I

INTRODUCTION

In this Part we explain the background to SACTRA's new terms of reference, and make some comments on the various ways in which roads and road transport interact with the environment.

CHAPTER 1
BACKGROUND TO THE REPORT

1.01 The decision whether to build a new trunk road and, if so, under what constraints, results from the weighing up of a number of different and sometimes sharply conflicting considerations. It has become the practice to divide them into two broad categories: "economic" and "environmental". As we shall show, both categories, and in particular the second, require considerable explanation. Between them, different schemes strike a different balance. The environmental benefits of building a by-pass around a village may outweigh the economic cost. The economic benefits of building a new road between two places may justify some unavoidable environmental damage.

1.02 The methods employed for assessing those factors which, under the current practice, are treated as "economic" have been established for many years. The economic assessment is based upon cost-benefit analysis, and in particular a specialised computer program known as COBA. Its aim is to provide decision-makers with an estimate of the economic costs and benefits over time of any new scheme relative to those which attach either to the continuing use of the existing road network ("Do-nothing"), or (more usually) to the use of the existing road network with such modifications as could in any event be expected to be carried out to it ("Do-minimum"). The operation of COBA is more fully described in Chapter 4 of this Report.

1.03 These economic data have to be set alongside many other effects which are not currently expressed in economic or monetary terms and are, therefore, outside the purview of COBA. They include many of the scheme's environmental and social effects, and they too form an important part of the appraisal. With a view to ensuring that they are properly considered alongside the economic data they are assembled into a framework, together with the COBA figures, in a form which is intended to show both to the public and to the decision-makers the full scope and importance of each of the impacts which a scheme might have.

1.04 This framework, which includes the COBA results, is prepared at various stages during the planning of a scheme and is presented finally for discussion, together with the more detailed evidence and reports relied upon in support of the scheme, at any Public Inquiry which may take place. Finally, all the material, including the framework, will be submitted to the Secretaries of State (who, in the case of trunk roads in England, are the Secretaries of State for Transport and the Environment) together with the Inspector's Report for final decision.

1.05 Guidance on the preparation of frameworks, and the appraisal of the environmental effects of road schemes, is contained in the Department of Transport's Manual of Environmental Appraisal (MEA). These guidelines are described in more detail in Chapter 6.

1.06 Cost-benefit analysis of one form or another has been used by the Department to assess trunk road investments ever since the 1960s. The COBA computer program itself was first developed in the 1970s. The current version, COBA 9, was issued in 1981 and has, on a number of occasions, been updated and amended in detail, most recently in 1989. New traffic forecasts were incorporated in 1989 and then re-based in 1990 to take account of the latest traffic statistics.

1.07 The MEA is based on procedures jointly developed by the Department and this Committee (SACTRA), and its predecessor the Advisory Committee on Trunk Road Assessment (ACTRA)[1]. The MEA was published in 1983 and relatively minor revisions were issued in 1988. Further revisions are under consideration.

1.08 In the intervening period scientific and general public awareness of the threats posed by many forms of human activity to the environment has been greatly enhanced, and the need to protect and conserve the environment now stands high in the order of political priorities. The Government's own strategy has been comprehensively set out in its White Paper "This Common Inheritance", Cm. 1200, published in September 1990. The European Community has laid down standards for assessing the environmental impact of all types of major development, including roads. At the same time, some commentators, most notably the London Environmental Economics Centre in its Report to the Department of the Environment, "Sustainable Development"[2], have suggested that the distinction between economic and environmental impacts may be false. In view of the passage of time and the important changes which have taken place this Committee was asked:

> "To review the Department's methods for assessing environmental costs and benefits, in particular (to consider) whether a greater degree of valuation is desirable, the appropriate scope and application of valuation and suitable methods for deriving monetary values".

1.09 We have interpreted our terms of reference broadly and flexibly (as the subject demands); and in this Report we present the review which we have been asked to carry out. We were able to consult a large number of bodies and individuals with particular expertise on the topic, who kindly submitted to us their views in writing. In some cases these were reinforced by considerable evidence and reasoned discussion. In addition, we held a one-day seminar on 14th September 1990, at which we were able to exchange views with and listen to many of those who had written to us. The names of those who have assisted are set out in Annex 1 to this Report. We wish to record our grateful thanks to them. The present membership of SACTRA and our full terms of reference are set out in Annex 2. We alone, of course, are responsible for the opinions which we express.

1.10 Our Report is arranged in the following sequence:

Part I
(Chapters 1 and 2): Background to the Report and a general view of the many and various ways in which roads and traffic affect the environment.

Part II
(Chapter 3): Description of the manner in which the planning of trunk road schemes evolves and passes through various stages of assessment and approval.

---

[1] Report of the Advisory Committee on Trunk Road Assessment, HMSO, October 1977 ("The ACTRA Report") under the Chairmanship of Sir George Leitch KCB, OBE.
[2] "Sustainable Development: The Implications of Sustainable Development for Resource Accounting, Project Appraisal and Integrative Environmental Policy", London Environmental Economics Centre, 1989 (authors: Professor David Pearce, Dr Anil Markandya, Dr Edward Barbier).

Part III
> (Chapters 4 to 8): Description of the way in which current methods for appraising the economic and environmental effects of road schemes have developed; summary of the views expressed on environmental appraisal by ACTRA and SACTRA in earlier Reports; and an account of the legal requirements relating to environmental assessment laid down by the European Community.

Part IV
> (Chapters 9 to 12): Discussion of the adequacy of current appraisal methods and recommendations for their improvement.

Part V
> (Chapters 13 to 15): Discussion of the extent to which a greater degree of monetary valuation of environmental effects is either feasible or desirable, and recommendations for the further development of the economic assessment of such effects.

Part VI
> (Chapter 16): Conclusions and Recommendations

CHAPTER 2
ROADS AND THE ENVIRONMENT

2.01 We give to the expression "the environment" the widest possible meaning. We take it to refer to every aspect of mankind's surroundings. It includes the physical resources of the planet; all animal and plant life; the earth, water and atmosphere. To that must be added the built environment: settlements and structures which serve and support human society and form part of our history and culture. Human activity itself is another component part. Industry, agriculture, sport and the recreational pursuits of some people become by their very presence part of the environment of other people. It might be said that "the environment" consists of everything which affects, collectively or individually, our health, comfort, safety or enjoyment of life.

2.02 Concern for the environment, however, is not exclusively centred on mankind. We also attach some intrinsic importance to the protection of the environment of other animal and plant species, for their own benefit, and not necessarily for any direct or indirect advantage which they might confer on us.

2.03 It is widely acknowledged that road building and road traffic have had a substantial impact upon the environment. Despite the many economic and social advantages which have resulted much of the impact has been detrimental. During the past thirty years, vehicle ownership has grown consistently (see Figure 1). The main cause is the continuing rise in real incomes. Another contributory factor has been migration away from city centres, which has made it easier to own and more desirable to use motor vehicles, and it is a continuing trend. The resultant rate of growth of the number of vehicles in the United Kingdom is comparable to that in most other European Community countries although, currently, the ownership rate (350 per 1000 people) is at about the European Community average. Further substantial growth is predicted by most forecasters.

2.04 The construction of roads, and their subsequent use by traffic, interact with the environment in countless ways. Some broad classification of the various types of impact is necessary to understand the different ways in which different impacts have to be measured and evaluated; to ascertain which environmental effects are and which are not relevant to the assessment of any highway proposal; and to provide decision-makers with a rational basis for comparative appraisal between competing options. We do however believe that excessive categorisation can result in bias. Hard and fast distinctions are difficult to justify. Many effects tend to lie in a continuum rather than within mutually exclusive categories.

2.05 The most frequently suggested dimension for classifying environmental impacts which emerged from the views expressed to us was that of scale. The argument runs as follows:

    a. Some of the environmental effects of roads and traffic can only be viewed in the broadest terms. Policies determined at the highest level of Government will determine the total number of miles of new road built and (to some extent, therefore) influence the total volume of traffic passing along that network. In building and maintaining these roads, we will consume many non-renewable

FIGURE 1
Increase in Car Ownership

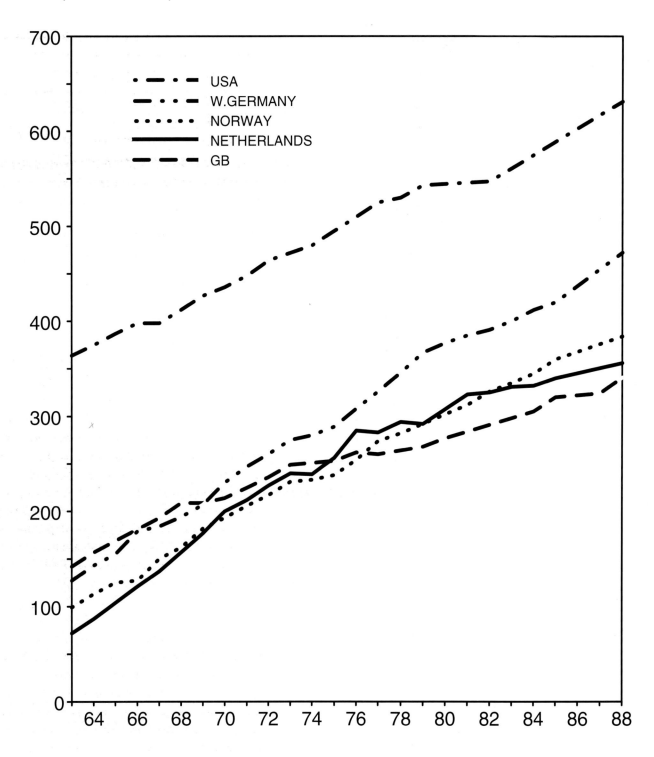

SOURCE : Transport Statistics Great Britain

natural resources, and permanently change the physical make-up of the land and the pattern of activities associated with it. Other non-renewable resources (notably fuels) will be consumed by traffic itself. Indeed, vehicle emissions are one of the main contributors to air-pollution. Traffic movement overall determines (for example) the quantity of vehicular emissions of carbon dioxide and oxides of nitrogen, which we now believe to be of global significance in the most literal sense.

b. Other effects may be more relevant to a particular part of the country as a result of a new route or corridor being developed. The most striking example is the change in the pattern of land use which is brought about by major road-building. Such changes will have a regional or at least a network-wide environmental significance, and are (incidentally) more difficult to predict or control than the more local impacts on the environment.

c. The balanced choice which has to be made between different options for some particular scheme or proposed road improvement, however, may be too fine to reflect many of these more general issues. For example, there may be virtually no difference between, say, Route A, Route B and the Do-minimum option so far as overall vehicle emissions are concerned. On the other hand, the relative nuisance suffered by local residents as a result of traffic noise, vibrations, visual intrusion or community severance are purely local effects which may be radically different in each of those three cases.

2.06 This consideration of environmental impacts at different levels of decision-making has much to recommend it. But, in practice, it requires a careful exercise of judgement. Take for example the destruction of a view by a viaduct. Is it just an agreeable outlook from one or two buildings that is being destroyed? It may instead be a local beauty spot or a fine sweep of countryside or coastline which forms part of our national heritage. If buildings are to be demolished, or the habitat of plants or animals to be uprooted, it is necessary to examine closely the subject-matter to judge at what stage consideration of that impact should be made and by whom. It is one thing to propose knocking down some buildings of no great architectural importance but quite another to demolish a historic monument, a listed building or part of a conservation area. If plant or animal life is being disturbed with the result that a rare or possibly unique species will be lost, that is likely to be a matter of general public concern. The same may not be true however if the species affected exist in abundance elsewhere. It may well be helpful to construct a scale or spectrum of effects, leading from global or national at one extreme to purely local at the other, but it will not always be easy to decide the exact location of any particular effect on that scale.

2.07 There is a further complicating factor, namely the cumulative effect of such decisions. The environmental damage caused by one individual scheme may be purely local, discrete and self-contained. But does it make a difference if similar damage is being replicated up and down the country by a number of other schemes? One topical example is that of trees and woodland. One road may cause the destruction of a wood which, taken by itself, might not be a serious loss. If, however, the Government is promoting concurrently one hundred road schemes, each one of which involves the loss of some woodland, then the impact of those schemes on the stock of trees may have to be looked at not simply in isolation, but in a national context too. In such cases, the total effect of the Government's roads policy nationally might be greater than the sum of effects of its constituent parts.

2.08 The environmental effects of road building and road traffic, across the full spectrum, cannot therefore all be reflected at one single level of decision making, and certainly not just at the level of scheme assessment. If all effects are to be taken into account in Government decisions, the appropriate environmental appraisal must underlie every stage in the hierarchy of decisions, from the making of national and regional policy downwards. Equally, it has to be recognised that different effects cannot be neatly compartmentalised, or allocated exclusively to one particular point in that hierarchy. As with all land-use planning, each level interacts with the other, and the decision process has to be an iterative one.

2.09 Examples are given in Figure 2 which illustrate the wide range of environmental effects which a road building decision can have. The table is also organised to show how different effects will arise at different stages on a "cradle-to-grave" progression from the formulation of policy (A) through construction (C) and operation (D) to the renewal and recycling of a road at the end of its life (F). Setting out the impacts in this way reminds us of the many and various resources which transport requires and of the manifold consequences for the environment of consuming these resources. To take just two examples, 34 million tonnes of road stone were excavated in 1989 from quarries around the UK[3] and no fewer than 23 million worn tyres are discarded by vehicle owners each year for which there is no adequate recycling process.

2.10 Another clear implication from Figure 2 is that so wide-ranging are the impacts that no one profession or branch of science is capable of identifying and assessing all the effects. This is as true at the local level as it is at the national or international level. It underlines the importance of multi-disciplinary cooperation in measuring and valuing environmental impacts in the most appropriate ways. We will have more to say on this later in the Report. There is a procedural implication to this also, in that the right forms of expertise need to be sought at each stage in the assessment and the appropriate degree of detail stipulated so as to inform the kind of decision being made in the most effective way.

2.11 There are other dimensions of importance; for instance, that of timescale. For the purposes of economic evaluation, a road is normally assumed to have a 30-year life. Thus the short-, medium-, and long-term impacts on the environment should be distinguished, so that the future environmental effects can be discounted in the same way as other benefits and costs. To the extent that traffic is expected to increase in future, one might expect environmental conditions to get worse. But they may, in practice, be offset by longer-term and system-wide effects of transport policy and continuing improvements in technology. Thus the time-varying impacts need to be considered in different ways, using the best technological forecasts available.

---

[3] "BACMI Statistical Year Book 1990", British Aggregate Construction Materials Industries.

FIGURE 2
Examples Covering the Wide Range of Environmental Effects of Trunk Road Development.

SCHEME - SPECIFIC "LOCAL"	CORRIDOR / PROGRAMME "REGIONAL"	STRATEGIC / POLICY "NATIONAL / GLOBAL"
(A) POLICY FORMATION		
(i) Energy	∗ Infrastructure required for fuel distribution and storage	∗ Pollution from transport / refining of crude oil ∗ Pollution from process of vehicle manufacture ∗ Implications for global warming (eg CO_2, CFCs) ∗ Implications for import / export of fuels / oil
(ii) Materials	∗ Roadstone / gravel resources / cement	∗ Mineral extraction and refining ∗ Implications for import / export of materials
(B) ROUTE SELECTION		
(i) Heritage and landscape ∗ Tree preservation ∗ Listed Buildings ∗ Archaeological sites ∗ Severance of farms	∗ Vulnerable landscape / AONBs ∗ Heritage buildings ∗ Archaeological areas	∗ Effects on National Parks ∗ Unique buildings and irreplaceable artefacts
(ii) Ecological disturbance ∗ SSSIs ∗ Sites of Biological interest	∗ Effects on feeding & breeding of species ∗ Special habitats	∗ Sites of international importance (eg RAMSARs National Nature reserves, SSSIs)
(iii) Planning considerations ∗ Pattern and scale of development ∗ Sterilisation of land ∗ Blight	∗ Site location / access ∗ Loss of countryside and open space ∗ Visual quality (structures)	

	SCHEME - SPECIFIC "LOCAL"	CORRIDOR / PROGRAMME "REGIONAL"	STRATEGIC / POLICY "NATIONAL / GLOBAL"
(C)	**CONSTRUCTION AND MAINTENANCE**		
(i)	**Site Location / Access**		
∗	Visual intrusion (structures)	∗ Damage to landscapes	
∗	Severance of existing roads		
∗	Loss of archaeological sites	∗ Loss of heritage buildings	
∗	Diverted traffic - noise - vibration - air pollution - visual intrusion (vehicles) - intimidation		
(ii)	**Construction traffic**		
∗	noise nuisance		
∗	vibration damage		
∗	dust		
∗	fumes		
∗	visual intrusion (vehicles)		
∗	intimidation		
(iii)	**Materials and Services**		
∗	Diversion of existing services	∗ Roadstone quarrying - noise - vibration - dust, fumes - visual intrusion (vehicles)	
(iv)	**Ecological disturbance**		
∗	Alteration of natural watercourses	∗ Intrusion in AONBs	∗ Effects on National Parks
∗	Damage to habitats	∗ Effects on feeding and breeding of species	∗ Loss of SSSIs
(D)	**OPERATION**		
(i)	**Traffic effects (petrol - engines)**		
∗	Noise nuisance	∗ Photochemical smog	∗ "Greenhouse" effect - CO_2 - CFCs
∗	Visual intrusion (vehicle)	∗ Effects on vegetation (eg CO_2, NO_x)	

SCHEME - SPECIFIC "LOCAL"	CORRIDOR / PROGRAMME "REGIONAL"	STRATEGIC / POLICY "NATIONAL / GLOBAL"

∗ Air pollution:
  - CO, ozone, lead
  - oxides of nitrogen
  - HCs aldehydes
  - evaporants
 	
∗ Deterioration of building facades
	∗ Induced traffic (due to redistribution and mode - shift):
	  - noise
	  - vibration
	  - air pollution
	  - visual intrusion (vehicle)
		∗ Energy implications of traffic growth

(ii) Traffic effects (diesel - engines		

∗ Noise nuisance
 	
∗ Vibration damage
 	
∗ Visual intrusion (vehicles)
 	
∗ Dirt / carbon dust
 	
∗ Air pollution:
  - oxides of nitrogen and sulphur
  - HCs, black smoke
  - benz - pyrene
  - aerosols
 	
∗ Deterioration of building facades
	∗ Photochemical smog
	
	∗ Effects on vegetation (eg CO_2, NO_x)
	
	∗ Induced traffic (due to redistribution and mode - shift):
	  - noise
	  - vibration
	  - air pollution
	  - visual intrusion (vehicle)
		∗ "Greenhouse" effect
		  - CO_2
		
		∗ Acidic rainfall
		  - oxides of sulphur
		
		∗ Energy implications of traffic growth

(iii) Hydrogeological effects		
∗ Chemical spillage ∗ Rubber detritus ∗ Oil leakage ∗ Contaminated surface run - off ∗ Soil pollution	∗ Chemical pollution of aquifers	
(E) DEVELOPMENT / LANDUSE CHANGES		
(i) Site locations and access		

∗ Loss of open space
 	
∗ Loss of archaeological sites
 	
∗ Visual intrusion (new buildings)
	∗ Intrusion in landscapes
	
	∗ Roadstone Quarrying
	  - Noise
	  - Vibration
	  - Dust, fumes
	  - Visual intrusion (vehicles)	

	SCHEME - SPECIFIC "LOCAL"	CORRIDOR / PROGRAMME "REGIONAL"	STRATEGIC / POLICY "NATIONAL / GLOBAL"
✶	Additional roads and traffic effects		
(ii)	**Traffic effects**		
✶	Generated traffic - Noise - Vibration - Air pollution - Visual intrusion (vehicles)	✶ Photochemical smog ✶ Effects on vegetation (eg CO_2, NO_x)	✶ "Greenhouse" effect - CO_2 ✶ Acidic rainfall - oxides of sulphur
(iii)	**Ecological disturbance**		
✶	Damage to habitats	✶ Effects on feeding and breeding of species	
(F)	**WASTE AND RECYCLING**		
(i)	**Highway and bridge reconstruction**		
✶	Demolition traffic: - noise - vibration - dust, fumes - visual intrusion (vehicles)	✶ Spoil heaps of old highway materials (visual intrusion)	
(ii)	**Spilled loads**		
✶	Litter / debris		
✶	Non - hazardous		
✶	HAZCHEM spillage: - Corrosive - Toxic - Flammable - Peroxide and organic - Radio - active	✶ Effects on animals / vegetation ✶ Chemical pollution of aquifers	
✶	Contaminated surface run off		✶ Radiological hazards
(iii)	**Vehicle and Component scrappage**	✶ Land - fill dumping of old tyres (fire hazard and visual intrusion) ✶ Scrap - yard heaps of old vehicles (visual intrusion) ✶ Incineration plants (air pollution) ✶ Land - sumps for old lubricating oils (chemical pollution of soil and aquifers)	✶ Energy input to recycling processes ✶ Acidic rainfall (due to incineration - oxides of sulphur

2.12 We can also divide impacts on human beings, either individually or collectively, into those which are perceived, and those which are not. Each category calls for its own method of assessment. Thus, some pollutants from traffic (for example lead) are absorbed by the human body without the persons concerned being aware of it. This process can occur over a long period. Ill effects may eventually be felt, but only when dangerous accumulations in the body have built up. Even then they may be of such a non-specific kind that the sufferer would not be able to attribute them to the real source of the problem. These unperceived effects have to be dealt with from the standpoint of public health. Their control, prevention and treatment depend upon clinical studies of their physiological effects and adequate safety levels need to be defined and judged against the public costs of achieving them.

2.13 The perceived effects, on the other hand, are those which are regarded by people as nuisances, whether or not they pose any clinical threat. Noise and black smoke, for example, are both perceived as annoying long before they actually endanger health. Thus, to treat these effects simply as public health hazards would be to disregard entirely the burden of the complaint which is made about them. Appraisal in these cases must be based on the measurement of subjective responses, correlated wherever possible with an objective measure of the effect itself. A good example of this is the adjustment of noise levels, expressed as A-weighted decibels, to allow for the subjective response to noise as a nuisance.

2.14 Figure 2 also demonstrates that while there are some effects which bear on people, whether as individuals or in a community, others lie remote from human well-being, for example they affect flora, fauna and habitat. Between these two extremes there is yet another spectrum of environmental characteristics which require appreciation as contributing to human welfare, and which yet have their own inherent value: soil, water, air, climate and landscape.

2.15 These general observations underlie the difficulty of the task we have been asked to undertake. An effective system of measurement and appraisal of traffic schemes must not only be able adequately to capture all the facets of the various environmental effects which are relevant to them, but also to structure them and give them coherence with policies and decisions made at all levels in the planning of our transportation system. The scope and variety is such that a wide range of different methods and procedures will be required to meet the task.

PART II

THE PLANNING OF TRUNK ROADS

In this Part we describe how trunk road schemes presently evolve and develop through successive stages of planning until they are approved by the Secretaries of State and are constructed.

CHAPTER 3
THE PLANNING OF TRUNK ROADS

3.01 To place our discussion in its proper context, we give a brief description of the process whereby road schemes find their way into the national roads programme, and pass through a succession of analyses, consultations and decisions, until they are finally built and opened to traffic. This Chapter covers much of the ground of Chapter 3 of SACTRA's Report "Urban Road Appraisal"[4] but focuses more closely on the point at which an environmental assessment of any kind is carried out.

3.02 The main stages in the planning and construction of a trunk road or road scheme are shown in Annex 3. Proposals result from an interaction between national policy, expressed in Government White Papers, and local initiatives which may come from the joint Regional Offices of the Departments of the Environment and Transport, MPs, local authorities or members of the public. The motives for schemes vary considerably. They may be the relief of local congestion, the reduction of accidents, the improvement of amenity or, in the case of a major route or corridor, the achievement of more general policy objectives (eg improving freight access to ports).

Identification Studies

3.03 When a need or problem has been clearly defined, it will lead in the first place to either a Route or a Scheme Identification Study, which will investigate a range of possible solutions. The purpose of such a study is to establish whether or not it is sensible to progress to the next stage of analysis. These studies will normally be handled in the Department's Regional Offices.

3.04 In some cases, a more strategic corridor study may be produced - ie a Route Identification Study. The purpose of a Route Identification Study is to investigate the consequences of upgrading a road over a long distance or of improving a network of roads. A current example is the Trans-Pennine Study, mentioned in the 1989 White Paper "Roads for Prosperity". There is no standard format for a Route Identification Study, but it will usually contain some broad environmental considerations, such as a recognition that the new line will pass through a certain range of hills or area of countryside. A Landscape Architect, who is either a member of the staff within the Regional Office or (more frequently nowadays) an independent consultant will contribute to the study. A Route Identification Study must take account of relevant national policies, such as the undertaking to avoid (as far as possible) any proposal to construct or upgrade a trunk road in a National Park. If a Route Identification Study suggests that study of the major route or corridor should progress, it will generate separate Scheme Identification Studies for individual parts of the route.

3.05 A Scheme Identification Study, which relates either to part of a Route or (as is more common) to an independent and less extensive site will be produced. It will include: (a) a definition of the area

---

[4] SACTRA 1986 Report "Urban Road Appraisal", HMSO, June 1986, under the Chairmanship of Professor T E H Williams, CBE.

in question with a map; (b) a note on any identifiable problems which the scheme is intended to solve, quantified so far as is possible; (c) a statement on the national objectives, if any, which would be served by the scheme; (d) a list of relevant local objectives; (e) a report on the views of the public, if known; (f) a description of all the options to be examined, and how they could be evaluated against the stated objectives; (g) an economic assessment; and (h) a statement on the further data required. This work is normally carried out by consultants and a Landscape Architect, as sub-consultant, may present initial findings on environmental matters such as existing landscape quality, heritage sites and the likely visual impact of any options under consideration. The Scheme Identification Study will contain a general statement of environmental effects, but is more likely to identify sensitive areas, and uncover environmental problems, than contain a detailed appraisal.

Entry into the Roads Programme

3.06 Once the Regional Office is satisfied that there is a case for more detailed study to proceed, it will be submitted to the Department's Headquarters and considered by Ministers for entry into the Roads Programme. Most of the studies which are approved are announced in a White Paper or Roads Report approximately every two years. A variety of data must be provided for this purpose, including some broad-brush economic assessment, and a statement of any significant environmental impacts perceived at that time. If a scheme is accepted into the programme it will, according to its priority, progress to the next stage, which ends with the presentation of options for Public Consultation. Entry into the Roads Programme does not imply a commitment to carry out a scheme. It is merely an announcement of the Government's intention to build, provided further studies show it to be justified, and eventually affordable.

Procedure after Entry into Roads Programme

3.07 After entry into the Roads Programme, schemes are subjected to more detailed analyses and modelling, which include traffic surveys, economic and environmental assessments, consultations in confidence with local authorities and other public bodies, and a report from the Landscape Advisory Committee. The relationship between the scheme and any relevant Structure or Local Plan prepared by the local planning authorities will then be explored. Reports include a Local Model Validation Report and a Technical Appraisal Report. The latter will include or be accompanied by the presentation of all the effects which are predicted at that stage, including environmental effects, in the first formal framework, as mentioned in paragraphs 1.03 and 1.04 above. Options will then be exhibited locally and be presented for Public Consultation. We comment in more detail below on the nature and extent of the environmental assessment which is carried out at this stage. We do, however, draw attention to the fact that, where a series of individual schemes has been generated by a Route Identification Study or (in the absence of a Route Identification Study) when a series of schemes taken altogether amount de facto to an improved route, detailed environmental assessments are always made but only of the individual component schemes. Corresponding assessments are not made at the aggregate level in relation to the overall route and its regional impact.

Preferred Routes or Schemes and Public Inquiry

3.08 Further analysis is carried out, and the studies are refined and drawn up in more detail, in the light of the opinions expressed at Public Consultation. The Secretary of State then chooses his Preferred Route. Draft Statutory Orders are prepared under the Highways Act 1980: Line Orders set out the line of the proposed road, Side Road Orders relate to any changes in access to or across the road, and Compulsory Purchase Orders define the land which the Department needs to acquire to build the scheme. In some cases, all three Orders are considered at the same Inquiry. In others, they may be considered at separate Inquiries. In straightforward cases, the Department is sometimes able to negotiate with persons directly affected by its scheme and an Inquiry can be dispensed with.

3.09 In virtually every case a formal Environmental Statement (see Chapter 8 below) must be published with the Orders and considered at the Inquiry with supporting data and information. The submitted material will also include the framework as mentioned in paragraph 1.04 above. The Inquiry is conducted by an Inspector. In England, the Inspector is appointed from a list maintained by the Lord Chancellor by the Secretaries of State for Transport and the Environment. They jointly consider his Report and decide whether the scheme should go ahead, either in its original form or with such amendments as they think fit, in the light of the Inspector's findings and recommendations.

3.10 Thus, it can be seen that a well established procedure exists in this country, within which all proposals for trunk road development can be studied, amended and exposed to scrutiny. It is in relation to this procedure that the adequacy (or otherwise) of methods for assessing the environmental consequences should be reviewed.

PART III

THE DEVELOPMENT OF CURRENT ASSESSMENT METHODS

In this Part of the Report we explain current appraisal methods, describe how they have developed over time, and summarise the earlier work of ACTRA and SACTRA in connection with environmental appraisal.

CHAPTER 4
CURRENT APPRAISAL METHODS:
TAM and COBA

4.01 Appraisal is carried out continuously while road schemes are under consideration, from the early development of options or "competing" proposals until the final design of an actual scheme. The two stages which are of especial importance are: first, when a number of schemes is presented for Public Consultation and, secondly, when the Secretary of State has selected his Preferred Route and it is submitted to Public Inquiry before an Inspector.

Traffic Appraisal and Traffic Forecasting

4.02 The exercise begins with a traffic appraisal. The full technicalities of this analysis are contained in the Department's Traffic Appraisal Manual (TAM). The amount, distribution and types of traffic using the existing network are analysed and its likely future growth is forecast. The forecasts are based upon the National Road Traffic Forecast (NRTF), which is prepared by the Department and is closely linked to Government forecasts of expected economic growth. The traffic forecasts used are of crucial importance, because they lie at the heart of the environmental as well as the economic appraisal. Because of the inherent uncertainty in forecasting, two forecasts are made: a "high growth case" (based on high economic growth and low fuel prices) and a "low growth case" (based on low economic growth and high fuel prices). From that analysis further forecasts, both "high growth" and "low growth", are made of the likely pattern of movement along the suggested new network, by re-assigning to the new network those movements which are likely to be transferred to it. We now turn to the methods currently used for economic evaluation.

The COBA Program

4.03 The economic costs and benefits likely to result from the proposed new network are compared with those of the existing network ("Do-nothing"), or of the existing network modified by such changes as would in any event occur ("Do-minimum"). For these purposes the cost-benefit evaluation computer program known as COBA is utilised[5]. Again, the predicted high and low growth in traffic have to be separately considered. The variables incorporated within the COBA program are discussed briefly in turn.

Valuation of Travelling Time

4.04 Journey times along links and delays at certain junctions are calculated both for the scheme and for the existing network, based upon the low and high forecasts referred to above. Different types of

---

[5] COBA is unsuitable for use on some congested urban networks and congested assignment packages (eg SATURN) are used instead and evaluated through the Department's URECA program to give results which are comparable to COBA.

journey-time are identified:

    a.   time spent on travel in the course of work ("working time"); and

    b.   time spent on travel for all other purposes including commuting to and from work ("non-working time").

Different monetary values are assigned to the travel-times for each type of journey. Working-time values are estimated by reference to average national wage-rates. Rates for car drivers and passengers are derived from the National Travel Survey; those for bus drivers and drivers and occupants of commercial vehicles come from the Department of Employment's New Earnings Survey. It is worthy of note that non-working time, which includes both commuting and leisure travel, is given a monetary value in COBA. This value is based upon a mixture of evidence including inference from actual traveller behaviour and preferences expressed by people in surveys and interviews. The standard appraisal value per minute for non-working time is currently about a quarter of that used for working time. On average non-working and working time benefits contribute about equally to the total benefits of trunk road schemes.

Valuation of Accidents

4.05   Road casualties are divided into two categories: fatal and non-fatal, with the non-fatal category being sub-divided into "slight" and "serious". Monetary-equivalent values are assigned to each type. The average value given to the avoidance of a fatality is derived from research using the "willingness to pay" approach. This research considered what people would be prepared to pay to reduce by a small amount the risk to them of fatal accidents. An additional allowance is included for the other economic (resource) costs, such as medical costs and damage to vehicles. Until 1988 the value was based on the "Human Capital" approach, which places a value on the gross contribution the victim would have made to the economy plus an allowance for pain, grief and suffering. Pending the results of current research, this is still the basis of valuation used for non-fatal injuries. Currently, therefore, the average value of preventing a fatality (in 1989 prices) is £608,580; of a serious casualty £18,450; and of a slight casualty £380. Damage to vehicles and property and certain police, insurance and administration costs are added to give the total values of avoiding different kinds of accident occurring on different kinds of road. The total "cost" of accidents on the "new" network is arrived at by multiplying the number of predicted accidents by the appropriate values. Predictions are expressed in terms of "Personal Injury Accidents per million vehicle-kilometres." Account is taken of the type of road, its length and daily average traffic flow, the number and types of junctions and the number of links. A comparison can then be made with the incidence of accidents likely to occur in the future on the existing network.

Vehicle Operating Costs

4.06   The COBA program takes account of five items of operating cost: fuel, oil, tyres, maintenance and depreciation. With the exception of the depreciation in value of private motor cars, these costs are, broadly-speaking, a function of distance and (to a lesser extent) speed of travel. Calculations are made of the changes in the distances travelled by all vehicles and on changes in average speeds on different links, according to the changes in flow. Comparative estimates, based upon the traffic forecasts, are made of these vehicle operating costs on each network (including the "Do-minimum").

Construction and Preparation Costs

4.07 The capital expenditure expected to be incurred by the new scheme will include: (a) the cost of all works contracts; (b) the cost of works to be carried out by other authorities such as British Rail and the providers of public utilities; (c) land and other compensation costs; and (d) administration and preparation costs, which include an allowance for the costs of the Public Consultation and the Public Inquiry. "Sunk" costs are excluded. For example, the costs incurred during Public Consultation are not retained in the economic appraisal at the time the Preferred Route is determined. All costs are measured in the constant prices of a given year, the so-called "Present Value Year", to which we refer below.

Cost of Maintenance

4.08 All road networks require recurring expenditure on maintenance. Some expenditure is not affected by traffic flow, for example the running costs of street-lighting, but much expenditure, such as periodic resurfacing and reconstruction, is directly related to it. Future maintenance may cause delays because of road works, and these may increase other user costs - delays, accidents and vehicle operating costs. Again, therefore, calculations of these, based on the predicted traffic flows, are made for all networks.

Discounting

4.09 The total amount of the recurring user costs described above (i.e. travelling time, accidents, vehicle operating costs and maintenance) is capitalised for each option over the period of construction and a 30-year life, discounted to the "Present Value Year", sometimes called the "Present Base Year". Discounting is undertaken (currently) at the annual rate of "8% real" to determine the Present Value of costs and benefits. Costs are thus expressed both for the proposed scheme and the "Do-nothing" or ("Do-minimum") network as single sums. Separate figures are given for "high growth" and "low growth" traffic forecasts in all cases. The difference between the net discounted user cost savings of the proposed scheme and its net discounted capital costs (including any maintenance savings) is referred to as the scheme's Net Present Value. This difference is calculated against the "Do-minimum" (or "Do-nothing") option. If the result of this comparison is a positive figure for Net Present Value, then the scheme is justified in economic terms. Usually, the Net Present Values calculated for all the feasible options are compared so as to identify the one which gives the best overall "value for money" - although a decision in favour of this option may be modified by a judgement concerning its relative environmental impact.

CHAPTER 5
ENVIRONMENTAL APPRAISAL:
THE ACTRA AND THE FIRST SACTRA REPORTS

5.01   In 1976, when the COBA method of cost-benefit analysis was just over 3 years old, ACTRA was appointed by the Secretary of State for Transport under the chairmanship of Sir George Leitch KCB, OBE. Its terms of reference were:

   a.   to comment on, and recommend any changes in, the Department's method of appraising trunk road schemes and their application, taking account both of economic and environmental factors, and of the extent to which these methods give a satisfactory basis for comparison with investment in alternative methods of transport; and

   b.   to review the Department's method of traffic forecasting, its application of the forecasts and to comment on the sensitivity of the forecasts to possible policy changes."

5.02   At the time ACTRA embarked upon its work, the Department's methods of appraisal of environmental and social factors were poorly developed. Some valuable but unpublished work had been carried out internally by a group of officials under the chairmanship of Mr J. Jefferson, then Deputy Director of the South West Road Construction Unit. The conclusion of their review, presented in March 1976, was that whilst it was impracticable to include environmental factors in cost-benefit analysis, a standard format should be adopted for their presentation. Indeed, the Department had already started to experiment with the practice before ACTRA began its investigations. The environmental and social factors which the Jefferson Report recommended for appraisal were:

   - land-take
   - noise
   - vibration
   - air pollution
   - visual effects
   - severance
   - accidents.

   With regard to land-take, the cost of acquisition was of course already included within COBA, but the property to be acquired was to be classified separately and its environmental effects assessed. Vibration and air pollution on the other hand were not regarded as significant for most schemes at that time. The role of the Landscape Advisory Committee in advising on the visual effects of road schemes was emphasised.

5.03   Many of those submitting evidence to ACTRA criticised the Department for attaching too much weight to road-user benefits, as quantified in COBA, at the expense of other factors. Little was known, outside the Department, of the work done by Jefferson. This evidence was summarised in

Chapter 13 of the ACTRA Report. It proceeded upon the assumption that there was a range of effects which were beyond the reach of quantification in any numerical sense, and its general tenor was that the Department paid scant attention to such effects, having no systematic method for dealing with them.

5.04 ACTRA itself considered that the current methods of assessment, based on COBA, were sound as far as they went, but were unbalanced, and that a shift in emphasis was called for. They said that it was "inadequate to rely simply on a checklist to comprehend environmental factors" and recommended that a comprehensive framework "relying on judgement" should be developed, in consultation "with all interested parties". Such a framework "should be employed from the earliest planning stages of a scheme"[6]. Their Conclusions and Recommendations[7] included the following:

- that the Department should adopt a framework for assessment as described above;

- that the assessment should take account of the effects of a scheme on five initial incidence groups: (a) road-users directly affected by the scheme; (b) non-road-users similarly affected; (c) those concerned with the "intrinsic value" of the area through which the scheme passes; (d) those indirectly affected by the scheme; and (e) the financing authority; and

- that the framework should be used to decide between options "on a basis of judgement, comprehending all factors whether valued in monetary terms or not."

5.05 As to the "non-economic component" of the assessment, detailed recommendations were made[8] on the development of techniques for describing and evaluating the importance of: (a) accidents (additionally to the monetary values assigned to them in the COBA); (b) the effects of schemes on pedestrians; (c) the loss of different types of buildings; (d) noise; (e) visual intrusion; (f) air-pollution; (g) disruption during construction; and (h) the effects on employment opportunities. In addition to accounting for the cost of its acquisition (at prevailing market values) in COBA, the amount of agricultural land taken was recommended to be reported separately in the framework, according to its grade. Likewise, the severance of agricultural land and of communities, in appropriate cases, was to be take into account. The effects of a scheme on the "intrinsic value" of important environmental assets such as National Parks, Conservation Areas and Historic Buildings, and landscape, ought (the Report concluded) to be explicitly identified.

5.06 ACTRA also recommended[9] that the Department should have available the services of a multi-disciplinary Standing Advisory Committee, charged with the following functions:

a. to advise on any significant changes proposed in appraisal or forecasting methods;

---

[6] ACTRA Report, Chapter 20, paragraphs 20.44 - 20.46.
[7] See ACTRA Report, Chapter 28 passim.
[8] ACTRA Report, Chapter 22.
[9] ACTRA Report, Chapter 29.

    b.   to initiate proposals for such changes; and

    c.   to recommend and make arrangements for studies in defined areas bringing in the appropriate expert advice.

In consequence of this recommendation, SACTRA was established by the Secretary of State in June 1978 again under the chairmanship of Sir George Leitch. One of the first topics which SACTRA was asked to examine was the development of the framework, which lay at the centre of the recommendations of ACTRA, and it reported on these matters in October 1979[10].

5.07   The First SACTRA Report contains three detailed worked examples of the assessment framework, referable to three different stages in the evolution of a hypothetical road scheme - Public Consultation, post-Public Consultation, and Public Inquiry. The principles upon which the frameworks are based are set out in Chapter 3 of that Report. The main guiding principles are: first, that no relevant effects should be excluded; and secondly that, while impacts should be quantified where this comes naturally, quantification is not an end in itself. No quantification should therefore be carried out unless there is a reliable methodology to support it.

5.08   The Report also addressed the question whether impacts which are quantified can be further expressed in money terms. In view of the relevance of this discussion to our present terms of reference, we reproduce some of the relevant paragraphs:

> "3.14 Where impacts can be quantified, it may be possible to evaluate them in a common unit, i.e. in terms of money. This is already done (in the COBA computer program) for travellers' benefits, where time savings, vehicle operating costs and savings in accidents are evaluated in pounds. If the effects of nuisance on (people in) houses, work places and places of recreation could be satisfactorily evaluated in pounds also, it would be possible to compare travellers' benefits and the effects on non-road users affected directly. Moreover, it would be possible to add the values to form a single index, thus extending the scope of cost-benefit analysis to encompass some environmental impacts..."

5.09   The First SACTRA Report went on to say:

> "3.15 We are not, however, convinced that this would be a welcome development for two reasons:
>
>   **a.**   the methodology depends on several assumptions about the way that people behave and how they choose to use their resources; it depends on the amounts involved being a small fraction of a

---

[10] "Trunk Road Proposals - A Comprehensive Framework for Appraisal", HMSO 1979 ("The First SACTRA Report").

person's resources, and on there being other facilities to which a person can switch so as to be able to maximise the enjoyment obtained from a fixed amount of money and time. It is therefore doubtful whether such techniques will apply to items such as housing, or to amenities which are regarded as unique.

b. there are objections to subsuming environmental and traffic effects into a single index. The impacts are of different types, are on different people and some will affect a small number of people greatly while others will affect a great number slightly. No uniform trade-off between them can be laid down, and there is no substitute for judgement in this area. Moreover, if these impacts were expressed in common units, there is a risk that the remaining impacts, which are often less tangible, would be given less weight in the evaluation than they deserve.

We therefore do not recommend that environmental benefits or disbenefits should be evaluated in money terms."

5.10　　The Report concluded[11] with the strong recommendation that the use of the framework should become standard practice in the Department. This has happened, and readers of the First SACTRA Report will see, in the worked examples, the early development of the practice which we go on to describe in the next Chapter. The framework which was developed as a result of the ACTRA Report is referred to in the following Chapters as "the Framework".

---

[11] First SACTRA Report, paragraph 6.3.

CHAPTER 6
CURRENT APPRAISAL METHODS: MEA, FRAMEWORKS AND COLOUR CODING

The MEA

6.01 The Manual of Environmental Appraisal (MEA) is produced by the Department of Transport as a "handbook of information" for use in the assessment of trunk road schemes. It is the product of the work undertaken by SACTRA and the Department of Transport. It contains three parts:

- first, the essential elements of the Framework and Departmental advice on environmental appraisal

- second, advice on the most appropriate method of assessing those specific environmental effects most likely to be important in route selection

- third, a list of major published sources on which the advice in the Manual is based.

Its contents have been prepared for trunk road schemes in England but they "might also apply to roads, particularly of an inter-urban nature, which are the responsibility of other highway authorities". (MEA Introduction, paragraph 8).

The Framework

6.02 The Framework is presented in a tabular form which is intended to draw together, in one document, the environmental and economic effects of a scheme in a manner which will facilitate the appraisal of a combination of quantified and unquantified effects produced by alternative methods. The focus at that time was upon inter-urban rather than intra-urban routes. The Manual states that:

"The information contained in the Framework is needed for two purposes:

- **a.** to provide members of the public with an appreciation of the important impacts of a proposed schemes and its alternatives; and

- **b.** as an aid to decision-making"

The way in which various factors can be brought together for decision-making is however the subject of another document - Advice Note 30/82 "The Choice Between Options for Trunk Road Schemes".

6.03 The Framework summarises "the main likely direct and indirect impacts on people of the alternative options for a proposed highway scheme". The Manual indicates that each option will have a separate column in the Framework and thus options are arranged comparatively within the one "balance-sheet". To ensure consistency between schemes the Manual specifies appraisal groups that are to be considered. There are six:

> Group 1, the effects on travellers
>
> Group 2, the effects on occupiers of property
>
> Group 3, the effects on users of facilities
>
> Group 4, the effects on policies for conserving and enhancing the area
>
> Group 5, the effects on policies for development and transport
>
> Group 6, financial effects.

6.04 In relation to each of these six Groups, the Framework brings together both the outputs of COBA and the appraisal of the environmental effects, so far as they are applicable to each of them. Thus, under Group 1 (Travellers), the travel-time savings and vehicle operating cost savings calculated by COBA will be set out; and under Group 6 (called Financial Effects but, in fact, including all monetised factors) the Net Present Values are presented based on "high-growth" and "low-growth" forecasts, for all proposed schemes, compared with the "Do-nothing" (or "Do-minimum") option. The Manual's title is therefore somewhat misleading, because it is not limited simply to environmental appraisal.

6.05 Frameworks are prepared both for the purpose of Public Consultation on a number of possible options, and for the purpose of the Public Inquiry into the Secretary of State's preferred route. At Annex 4 and 5, we reproduce the two types of framework suggested in the Manual to be appropriate to each stage.

The Environmental Appraisal

6.06 The second and more substantial part of the Manual is concerned with detailed environmental appraisal. Technical guidance is given on the appropriate method of measuring and reporting on eleven distinct effects. The reports which are to be drawn up form part of the documentary support for the presentation of different options at the Consultation stage. These are incorporated into the Department's case at the Public Inquiry. The main points are summarised for the purposes of the Framework and distributed as appropriate through the various Groups (in effect Groups 1 - 5).

6.07 The general thrust of MEA is to concentrate on effects which are at the local end of the spectrum and directly affect people, rather than the environment as a whole. Thus, Section 1 deals with Traffic Noise and Section 2 with Visual Impact (distinguishing visual obstruction from visual intrusion). Section 3 is concerned with Air Pollution, but only in the very local sense and, in accordance with advice previously given by ACTRA and SACTRA, is only investigated "where it is a particular problem". Section 4 (Community Severance), Section 8 (Disruption due to Construction) and Section 10 (View from the Road) also fall into this category.

6.08 Wider issues are raised under Sections 5 (Effects on Agriculture), 6 (Heritage and Conservation Areas) and 7 (Ecological Impact). In each of these cases, assessors are encouraged to consider impacts in a regional and national context. However, it is difficult to say that these wider considerations fit comfortably into the Framework, because the most obvious Groups to which they would belong - Group 4 (policies for conserving and enhancing the area) and Group 5 (policies for development and transport) - appear to be confined in scope.

6.09 The other two Sections - 9 (Pedestrians and Cyclists) and 11 (Driver Stress) - are concerned with the general environmental effects of the various schemes on special classes of travellers forming part of Group 1 in the Framework. Furthermore, they seem to be of a less "environmental" character than the other Sections.

Colour Coding

6.10 We also draw attention to an interesting method of classification which has been developed informally, and for purely administrative purposes, by the Department of Transport and the Treasury. The system is known as "colour-coding". It is used to give a rough and ready indicator of the environmental quality of a scheme across a green-yellow-red spectrum. We emphasise that it is not used for any purposes connected with statutory decision-making. A more detailed description of colour coding and its purpose is contained in Annex 6.

CHAPTER 7
URBAN ROAD APPRAISAL: THE SACTRA 1986 REPORT

7.01 The work carried out by ACTRA, SACTRA and the Department, which has been described in the preceding chapters, concentrated on the appraisal of inter-urban trunk road schemes. In July 1984, the Secretary of State invited SACTRA to review the Department's methods for assessing the environmental, economic and other effects of urban road improvements. SACTRA, under the Chairmanship of Professor T E H Williams CBE, presented its Report in June 1986. That Report was published in October 1986[12], together with the Government's considered Response to it.

7.02 The 1986 Report recommended a number of radical changes to existing procedures which are relevant to our present terms of reference, the majority of which were accepted by the Government in its Response. We set out a brief summary of the recommendations of especial importance to our present review. They are:

- that schemes should be assessed by reference to the stated national and local policy objectives which they are to fulfil;

- that road and public transport problems should be identified by reference to those objectives;

- that a range of feasible options should be developed which would satisfy the objectives; and

- that a traffic appraisal, economic evaluation and assessment of environmental and social impacts should be carried out in every case.

7.03 In the formulation of objectives, SACTRA stressed the need for a closer integration of the systems of planning roads and of other town and country planning considerations, at national and local levels. In the development of options, a strong call was made for greater flexibility of approach in the initial stages of the process, and for greater public participation at that stage. At present, as we have pointed out in Chapter 3, formal Public Consultation does not take place until after a scheme has entered the Roads Programme, by which time a great deal of technical work will have been carried out in defining the options in some detail. For complex cases, SACTRA recommended a Public Hearing at the very first stage. At such a Hearing, options would be tested against the objectives, and in the light of the identified problems which they were intended to solve.

---

[12] SACTRA 1986 Report "Urban Road Appraisal", HMSO, October 1986.

7.04 The principles of assessment, which had been proposed by ACTRA and were then fully operational, were in general endorsed, with necessary amendments made in the light of the special problems of urban roads. Changes were recommended to current techniques for traffic forecasting and economic evaluation (see our Chapter 4) with a view in particular to improving the quality of the material obtainable from COBA. None of these recommendations calls for further comment here.

7.05 Chapter 11 of the SACTRA 1986 Report contains a detailed consideration of the techniques for the assessment of environmental and social impacts. The Committee examined the advice given in MEA "and found its coverage and methods generally sound, in principle, but in need of some changes of emphasis to make it more suitable for application to the full range of urban schemes"[13]. The differences in emphasis which SACTRA recommended were set out in Figure 11.1 of the 1986 Report, and we reproduce it here as our Figure 3. The MEA was also found to be in need of some updating. Moreover, the establishment of a Townscape Advisory Committee as the urban equivalent of the Landscape Advisory Committee was recommended. The suggestion that Ecological Impact was "rarely important" might not, in the current climate of opinion, be universally accepted.

7.06 The Government, in its Response, accepted most of these recommendations. It accepted in principle the recommendation to include, in the environmental assessment, night-time noise, community severance and the cost of blight, subject to the outcome of research. It agreed that accident numbers should be detailed separately according to their severity, as well as being incorporated in the cost-benefit analysis. It also agreed that air quality reports were to be obtained only where effects were clearly identified as a problem; and that "view from the road" and "driver-stress" should not be considered as impacts in urban areas, unless (in the Government's view) there was likely to be a significant difference between options. The Government was unwilling to set up a new Townscape Advisory Committee, but undertook to modify the composition and terms of reference of the Landscape Advisory Committee.

7.07 On the question of principle with which we are now faced, the 1986 Report did not call into question the earlier opinion of ACTRA, namely that the assessment of environmental and social impacts was a matter of professional judgement (aided by measurement wherever possible) rather than of quantification and translation into monetary-equivalent values. Paragraph 12.6 of the 1986 Report contains this statement:

> "We do not recommend the use of impacts which are quantifiable as proxies for other effects which are not quantifiable or which are too difficult to measure. Impacts which cannot be quantified must be assessed on the basis of judgement, not left out. It is then equally important that the basis for these judgements is stated explicitly in the supplementary information which will support the assessment."

Systems of "scoring" or "weighting" were rejected, and support was given to the current approach of using a combination of cost-benefit analysis and judgement to arrive at decisions[14].

---

[13] SACTRA 1986 Report, Chapter 11, paragraph 11.4.
[14] SACTRA 1986 Report, paragraphs 12.14 and 12.15 and Recommendations 13.60 and 13.66.

FIGURE 3
Environmental and Social Impacts:
Summary of Changes Proposed for the Assessment of Urban Main Road Schemes

TYPE OF IMPACT LISTED IN THE MANUAL OF ENVIRONMENTAL APPRAISAL	CHANGES PROPOSED FOR ASSESSMENT OF URBAN MAIN ROADS
Traffic Noise	important; add night - time noise where relevant
Visual Impact	include with consideration of townscape
Air Pollution	can be perceived to be important
Community Severance	important; further research needed
Effects on Agriculture	rarely important
Conservation Areas	include with consideration of townscape and site - specific impacts
Ecological Impact	rarely important
Disruption due to Construction	important
Accidents	important
View from the Road	omit
Driver Stress	omit
Social Consequences of Blight	add description where relevant
Development Potential	add description where relevant

SOURCE : Urban Road Appraisal, The Standing Advisory Committee on Trunk Road Assessment, HMSO, June 1986

7.08 Finally, on the question of the presentation of assessments, the Report recommended, in effect, the replacement of the Framework method - which experience had shown to be unsatisfactory in some respects and inflexible - by a new format to be called an Assessment Summary Report. This Report would include five sections:

      a.   objectives and problems;

      b.   opinions and consultations;

      c.   traffic appraisal;

      d.   economic evaluation; and

      e.   environmental and social impacts.

Such a Report would conclude with a statement justifying the Department's preferred option. The Government also accepted this recommendation in principle, which had wider application beyond that of urban roads. Before issuing guidance on it, the Department undertook to experiment with it on one or two of its schemes, to gain experience. It has done so in the London Assessment Studies and elsewhere, although the recommendation has not yet formally been implemented.

CHAPTER 8
ENVIRONMENTAL ASSESSMENT UNDER EC DIRECTIVE 85/337

8.01 The Council of the European Community (EC) adopted its Directive (85/337) on "The assessment of the effects of certain public and private projects on the environment" on 27 June 1985. It came into force in July 1988. The effect of the Directive is to require an environmental assessment of certain types of major development to be carried out, and made available for public discussion, before development consent is granted.

8.02 So far as trunk roads are concerned, the EC Directive was given legal effect in England, by means of amendments to the Highways Act 1980[15]. The Department of Transport's Departmental Standard HD 18/88 explains how the Directive and the provisions of the Act are to be followed in practice.

8.03 For some types of development, the requirement for an environmental assessment is directly imposed by the EC Directive itself, and is mandatory. Those projects are referred to in Article 4.1 and Annex 1 of the Directive. They include the construction of motorways and other "express roads". The Directive defines "express roads" by reference to the terms of a European Agreement on roads made in 1975: "Roads reserved for automobile traffic, accessible only from interchanges or control junctions and on which, in particular, stopping and parking are prohibited". These roads equate to "Special Roads" under the Highways Act 1980.

8.04 For other types of development, the requirements only come into play if and so far as individual Member States of the European Community so decide. These projects are referred to in Article 4.2 of the Directive. They include the construction of all other types of road. The Government has decided that the following additional types of road scheme should be the subject of environmental assessment under the Directive:

- all new roads over 10 km in length;

- other new roads over 1 km in length which pass:

    a. through or within 100 m of
        i. a National Park
        ii. an SSSI
        iii. a conservation area
        iv. a Nature Reserve within the meaning of the National Parks Access to the

---

[15] Highways Act 1980, Part V A, introduced by SI 1988 No 1241.

Countryside Act 1949; or

    b.   through an urban area where 1500 or more dwellings lie within 100 m of the centre line of the proposed road;

- motorway and other road improvements which are likely to have a significant effect on the environment; and

- other new roads not within the above definitions if they are likely to have a significant impact.

For all practical purposes, therefore, the combined effect of these provisions is that virtually all road schemes have to be assessed in accordance with the Directive.

8.05 Article 3 of the Directive requires the assessment to identify, describe and assess in appropriate terms the direct and indirect effects of the project on the following:

- human beings, fauna and flora;

- soil, water, air, climate and the landscape;

- the interaction between the above two groups of factors; and

- material assets and the cultural heritage.

The wide terms of the Directive seem to invite an environmental appraisal in the fullest sense, which is not limited to local or even regional effects.

8.06 Article 5 and Annex 3 of the Directive expand upon Article 3, but some discretion is given to Member States as to how much detail the assessment should contain. It must at least: (a) describe the scheme and its site; (b) state the measures proposed to mitigate adverse environmental impact; (c) give the data required to identify and assess the main effects which the project is likely to have on the environment; and (d) contain a non-technical summary.

8.07 There are some additional matters which may, where appropriate, be included. Two in particular should be mentioned here. The first is an outline of the main alternatives studied by "the developer" and an indication of the main reason for his choice, including environmental effects (Annex III Clause 2). The second is a description of the likely significant effects on the environment resulting from: (a) the existence of the project; (b) the use of natural resources; (c) the emission of pollutants; and (d) the creation of nuisances and the elimination of waste. The developer must also give a description of the forecasting method used to assess these effects (Annex III Clause 4).

8.08 These requirements are laid down not merely for the benefit of decision-makers, but also to allow other public bodies and the public itself to state their opinions. Any other Member State likely to be affected by the project must also be notified and given the opportunity to express a view. No decision on a scheme can be made without taking into account the view so expressed.

8.09 The Department of Transport has had to adapt its own procedures for scheme approval to conform with the requirements of the Directive. In the Departmental Standard, the document containing the required material is called the "Environmental Statement". There is no requirement that, at the earlier Public Consultation Stage, when options are presented for discussion, there should be Environmental Statements presented for each of the schemes proposed at that time. The effect of this (and it appears to accord exactly with the terms of the Directive) is that, at the earlier Public Consultation stage, when options are presented for discussion, none of the proposed schemes is accompanied by an Environmental Statement in this form. The same is true of the "Do-nothing" (or "Do-minimum") option. If, at that earlier stage, any other options presented had environmental effects significantly different from the Published Scheme, the Environmental Statement which accompanies the finally preferred scheme must, in summary form, describe the alternatives and give reasons for making the final choice.

8.10 The required description of the scheme, and the measures adopted to mitigate any adverse environmental effects, go into the Environmental Statement itself, which to this extent is a separate document. The discussion of the alternatives considered at the Public Consultation, the reasons for choice of the Published Scheme and the non-technical summary are also found in the Environmental Statement. For the provision of data on environmental effects, the Department relies upon the contents of the appraisal Framework derived from MEA. The Standard does not require separate entries for any of the matters specified in Clause 4 of Annex III (see paragraph 8.07 above).

8.11 These procedures may be compared with those applicable to development projects which require planning permission under the town and country planning legislation. In those cases, the Environmental Statement must be submitted to the local authority with the application for permission and its existence must be advertised. It must be available for inspection, and for sale in reasonable quantities to the public. Since it is the only document produced by the developer which deals with environmental matters, it must be comprehensive, and all the environmental issues must be addressed within the four corners of that document. In the case of roads, the Environmental Statement arrives at the latest stage in the decision-making process. It may be argued that the publication of the Preferred Route is in many respects analogous to an application for planning permission, but, as we have shown, this event in the case of trunk roads will have been preceded by a series of earlier public decisions which do not have an exact counterpart under the town and country planning system. The Framework and its supporting documents will have been worked up during those earlier stages without an Environmental Statement and the Environmental Statement, when it arrives, is an adjunct to them.

PART IV

DISCUSSION OF CURRENT PRACTICE

In this Part of the Report, we consider the adequacy of current appraisal methods, in the light of the evidence which we have received, and make recommendations for their improvement.

CHAPTER 9
WHAT IS GOOD PRACTICE?

9.01 The body of material which we have summarised in Part III above contains a number of themes, not all of which are easy to reconcile. We have however endeavoured to distil from them practical advice as to how, in various ways, current practice might be improved. We started by examining first principles and tried to identify the essential features which any effective system of environmental appraisal must possess.

9.02 In Chapter 2 we drew attention to the great diversity of the environmental effects produced by roads and traffic. We pointed out that there is an ever-growing awareness of their nature and scope and that a fully effective appraisal system must be one which properly reflects all the environmental impacts which are known to occur, and not just some only, ignoring others.

9.03 The SACTRA 1986 Report contained the strong recommendation that the planning of trunk and other main road schemes in urban areas should be guided by explicitly-stated policy objectives. These objectives would then provide a series of reference points, against which different schemes could be assessed. In relation to each option, the question would be: to what extent does it promote or impede each of the stated objectives? In its Response to the Report, the Government accepted the recommendation. It has been strongly urged on us that this philosophy is particularly appropriate to environmental assessment. We endorse that view.

9.04 One point of particular importance has recurred continuously in our discussions, and in the evidence and advice we have received from others, namely the linkage between some at least of the effects of a particular scheme and regional, national and possibly global environmental interests. In Chapter 2, we gave some examples of the more obvious ways in which a single scheme might have a wider significance: eg damage to an important landscape, a unique habitat, historic sites or buildings. We can reinforce the point with further and more elaborate examples:

    a. New roads are often built and other improvements made to relieve a congested network. The result may be that some additional traffic is attracted to the network which now offers a higher level of service than would otherwise be the case. Since the overall quantity of traffic-demand (vehicle-kilometres) determines the level of air pollution due to vehicular emissions, the environmental assessment of any given local scheme should (on the face of it) be able to give an account to central policy-makers on the extent to which that scheme will exacerbate or relieve a national problem. But there are two difficulties. First, the differential effects between options for the scheme may be difficult to measure at the local level - it is the quality and cost of using the network as a whole which determines the volume of vehicle kilometres. Furthermore, unless the induced traffic is properly estimated, the indication of the effect could even be perverse.

    b. Similarly perverse results can arise at the scheme level due to the effect that the scheme has on traffic speeds, which are related in a complex way to vehicle emissions. The relief of congestion by measures that increase speed and smooth the flow of traffic will reduce specific fuel consumption, as well as make savings in travel-time. As speeds increase further, of course, specific fuel consumption will begin to rise again due to aerodynamic drag. In other cases, environmental benefits may be produced by reductions in speed, as in the case of traffic calming policies. Thus, depending on the nature of the scheme itself, overall vehicle-emissions may either rise or fall.

c. We have made the point earlier that transport systems and land-use also interact in a dynamic way. New development can be guided by development control to locations which may reduce the overall need for car journeys and encourage people to use more energy-efficient means of transport, such as public transport, walking or cycling. Conversely, transport can be planned to link up complementary developments with the same object in view.

These three examples underline further the need for a strategic level of assessment as reliance on scheme-appraisal alone may not give the full picture.

9.05 There is common ground that some way should be found to allow the agreed importance of all these factors to be given their due weight in the formal procedures which influence how much new road-building there should be, where it should be located, what design standards should be applied and the consequences of these and other policy decisions on the quantity and characteristics of traffic. This applies both to negative factors (how to stop making the environment worse) and positive ones (how to make it better). Assessment must be adequate:

- in geographical extent (not just in the immediate vicinity of the new road);

- in timescale (longer term as well as immediate effects);

- in its presentation of all alternatives (not just alternative routes, but also modes and even policies);
and

- in its consideration of interactions that is to say the combined and cumulative impacts of several schemes and policies.

9.06 Our views on the environmental assessment system that we need may therefore be expressed in these terms:

- First, it must derive from explicitly stated policy objectives. Its logic must be consistent with those objectives, and it must provide information concerning the degree to which the objectives are fulfilled by different schemes (including "Do-nothing" or "Do-minimum") from which a choice is being made.

- Secondly, it should ensure that every relevant environmental effect is identified and that all significant (or potentially significant) effects are fully and accurately described and measured in appropriate terms.

- Thirdly, it should be based on real evidence and the highest possible standard of scientific and technical work.

- Fourthly, it should be practical. The resources and time necessary to operate it should not be out of proportion to the importance of the issue involved, or greater than the cost of the outcome.

- Fifthly, it should be capable of being applied consistently between different alternatives, and between decisions taken at different times in respect of other proposals.

- Sixthly, it should be acceptable both to professional people and to the general public. To this end, the value judgements and technical assumptions made, and the methods used, should be explicit, open to scrutiny, challenge and possible revision.

- Finally, the language in which the appraisal is set out should be clear and non-technical, so that the basis upon which decisions are taken is open and capable of being generally understood.

CHAPTER 10
OBJECTIVES-LED ENVIRONMENTAL ASSESSMENT

10.01   In this Chapter, we consider the implementation of the first of the principles set out in paragraph 9.06 above. This is appropriate in view of its relevance to the recommendations of the SACTRA 1986 Report.

10.02   In the strategic development of the nation's roads, great emphasis has been placed upon the contribution which they make to our economic prosperity. Far less attention has been paid, it seems to us, at the strategic level, to their overall effect on the environment. This is not to say that environmental issues have been ignored. The development of by-passes, for example, has been strongly motivated by local environmental considerations, and there is, as we have shown in Chapter 6, a well-established practice for evaluating the environmental effects of a scheme in its local context. What has not occurred to the same degree is an examination of the environmental consequences of roads policy in aggregate. The absence of a strongly-stated central view has inevitably circumscribed the manner in which the appraisal of individual schemes has been carried out. Furthermore, it has left objectors to some schemes with a feeling that, within the system, there is no proper forum for the public debate of more strategic or global environmental questions.

10.03   Some change of emphasis could be seen in the Government's White Paper "Policy for Roads in England: 1987"[16]. Section 5 contains a lengthy statement of the Government's policies towards environmental issues. The key themes can be summarised, as follows:

- environmental improvement itself will be a major justification for new schemes;

- wherever possible new roads will be kept away from Areas of Outstanding Natural Beauty, SSSIs, National Parks and land owned by the National Trust;

- traffic management schemes, in particular in relation to parking, are to be promoted;

- the exact location and the detailed design of new roads will pay increasing regard to environmental effects;

- measures will be taken to protect wildlife and archaeological sites;

- wider consultation on environmental issues will take place; and

---

[16] Cm. 125, HMSO April 1987.

the special problems of lorries are to be tackled by lorry management schemes and improved vehicle design[17].

Even this statement, however, did not reveal any policy towards what are now regarded as fairly fundamental questions, such as the overall effects of vehicle emissions or the depletion of non-renewable natural resources.

10.04 The 1990 White Paper[18] represents a considerable advance on this position. A great deal of attention is paid to the effect of the road programme on (for example) carbon dioxide (CO_2) emissions (paragraphs 5.55 and 5.56), the protection of the countryside (paragraphs 7.37 and 7.38) and traffic management schemes (paragraphs 8.12 - 8.16). The Chapters on Air, Water and Noise (Chapters 11, 12 and 16) have important implications for national transportation planning.

10.05 Paragraphs 18.6 of the 1990 White Paper, contains this statement:

"During the preparation of this White Paper, each Government Department has reviewed the environmental implications of its existing policies and considered changes to them. Where this has resulted in proposals for change, these are included in the relevant chapters. To build on that exercise, and to ensure that the environmental implications of decisions are fully considered beforehand, the Government has carried out a review of the way in which the costs and benefits of environmental issues are assessed within Government. The review has looked at the range of analysis which is available on environmental costs and benefits, recognising the need for an integrated approach which takes account of all the consequences of a measure for the environment, favourable and unfavourable. It has concluded that there is scope for a more systematic approach within Government to the appraisal of the environmental costs and benefits before decisions are taken. The Government has therefore set work in hand to produce guidelines for policy appraisal where there are significant implications for the environment".

These guidelines are contained in a recent publication entitled "Policy Appraisal and the Environment", produced by the Department of the Environment[19]. We are encouraged that the publication is at many points consistent with our own thinking.

Setting Objectives and Targets

10.06 Implementation starts with the Government, at its highest level of policy-making, setting objectives or (where appropriate) specific targets in relation to the environmental impacts of the most concern, and establishing procedures for monitoring and reporting performance in the progress towards their achievement.

---

[17] Further policy statments on lorries are to be found in the White Paper "Lorries, People and the Environment" Cm. 8439, HMSO 1983.
[18] "This Common Inheritance" Cm. 1200, HMSO September 1990.
[19] "Policy Appraisal and the Environment", Department of the Environment, published by HMSO, 1991.

10.07 Clearly, it is not for us to say what those policies ought to be - but we can give some examples of the cases in which central guidance might be considered appropriate. Macro-environmental issues, such as atmospheric CO_2 and the use of non-renewable physical resources, for example fossil fuels and construction materials, can only be handled (if they are to be considered at all) at the transport policy level. Any national policies towards land-take and land-use, the conservation of species and habitats, the preservation of heritage and other assets of national importance should also be made explicit. Alternative transport strategies, including proposals for investment as well as changes in pricing, taxation and regulation can only be appraised against strategic environmental objectives and constraints. Transport strategies such as these should then be set within wider planning strategies dealing with the use and management of land and the state of the environment, at each of the levels from national to local.

10.08 Likewise, the procedures for the initiation, design and approval of trunk roads should relate in both timing and content to those for land-use and transportation planning. If due consideration of these important matters is not given at the outset, they are likely to be lost altogether when the somewhat narrower choices between different schemes at a more local level have to be made. This implies some formal assessment of environmental impacts at the regional level of a kind which we propose in the following Chapter 11.

10.09 Then, at the regional as well as the local level, detailed and specific environmental policies and objectives will come into play. Transportation planning, like all land-use planning, is an iterative process. As development proposals become successively more refined, they become more sensitive to local environmental concerns. The Departments of the Environment and Transport aim to ensure that strategic, regional and local planning objectives are, so far as possible, and where they overlap, properly coordinated. It is our view that the appraisal of individual road schemes will achieve a proper focus only if is it carried out within a framework of national, regional and local environmental policies and objectives working in combination.

10.10 On this approach, it follows that the early stages of the planning of trunk roads, up to and including both the Route Identification and the Scheme Identification Stages, are of great importance in the appraisal process. To this end, the procedures now followed by Regional Offices at the Route Identification Stage need to be codified more thoroughly. No scheme should be admitted into the Roads Programme until its performance against these strategic objectives and constraints has been evaluated and reported in outline. Similarly, where regional and more local environmental objectives have been adopted by local planning authorities which are relevant to roads, routes and schemes should be tested against these objectives also, at the Identification Stage. The results of this early appraisal should be available at the Public Consultation and Public Inquiry stages, so that the public may be satisfied that these matters have been fully taken into account, and to allow the results to be subjected to public scrutiny.

10.11 After a scheme has entered the Roads Programme it is likely that many of the "strategic" environmental questions will have been addressed, and the remainder of the process will concentrate, as it does at present, on the more regional and local effects which (in part) govern the choice between options. But that will not always be the case, and even at that stage room should prudently be left within the appraisal process to ensure that differences between schemes which significantly affect national objectives are properly picked up.

CHAPTER 11
THE PRESENTATION OF ENVIRONMENTAL ASSESSMENTS AND ENVIRONMENTAL STATEMENTS

The Timing of Assessment

11.01 In Chapter 10 we have emphasised that environmental assessment should be a continuing process throughout a scheme's development, and should commence at the earliest stage in the formulation of policy. The only relevant statutory requirement, derived from the EC Directive and the Highways Act 1980, is to present an Environmental Statement at a relatively late stage when Line Orders are published. Good practice, in our view, demands that formal assessments should be presented well in advance of that.

11.02 These interim presentations should successively deal with (i) the wider national and regional issues; (ii) the corridor within which route options are evolved; and (iii) the alternative scheme options themselves. The assessment reports should become increasingly detailed in coverage, such that the environmental assessment of a route-option should be readily developable into an Environmental Statement if that option finally goes forward to Public Inquiry.

11.03 We envisage, therefore, reporting on environmental assessment at the following stages:

    a. entry into the Roads Programme;

    b. Public Consultation on the options; and

    c. the publication of Line Orders for the Secretary of State's Preferred Route (for Public Inquiry if one is to be held). This is the point at which the Environmental Statement required by statute must be produced.

11.04 In each case, the assessment report will be a free-standing document. For the purpose of distinguishing between the statutory obligation to produce an Environmental Statement and the earlier non-statutory reporting advocated above, we propose that the interim reports are identified as follows:

    a. Environmental Assessment Report - Roads Programme Entry

    b. Environmental Assessment Report - Route-Options.

11.05 We consider that the "Environmental Assessment Report - Roads Programme Entry" must be available to those who subsequently undertake the principal work of environmental assessment and a summary of its main contents will be reported in the "Environmental Assessment Report - Route-Options". This latter report will be a document which is available to the public for purposes of public information and consultation and will be available for discussion at the Public Inquiry. With respect to it, and to the statutory Environmental Statement which is produced at Stage c (see paragraph 11.03), we suggest that copies be made available to those who are most directly affected by the proposals but that other interested parties are able to purchase copies of these documents on the basis of its price being related to cost recovery.

The Statutory Requirements for Environmental Statements

11.06 In paragraphs 8.05 - 8.07 we summarised the relevant provisions of Articles 3 and 5 and Annex 3 of the EC Directive and emphasised the breadth of their scope. Every significant effect, direct and indirect, must be reported. There is no bias in the discussion towards any particular point on the global - national - regional - local spectrum. The fact that, in appropriate cases, other Member States of the European Community must be given the opportunity to make their comments on an Environmental Statement produced under the Directive is a strong indication of its purpose. Properly designed, an Environmental Statement of this type is, in our opinion, the flexible document best equipped to meet the several criteria set out in paragraph 9.06.

11.07 The UK legislation[20] requires the Secretary of State to publish "a statement containing the information referred to in Annex III to the Directive to the extent that he considers:-

    a. that it is relevant to the specific characteristics of the project and of the environmental features likely to be affected by it; and

    b. that (having regard in particular to current knowledge and methods of assessment) the information may reasonably be gathered,

including at least-

    (i) a description of the project comprising information on the site, design and size of the project;

    (ii) a description of the measures envisaged in order to avoid, reduce and, if possible, remedy significant adverse effects;

    (iii) the data required to identify and assess the main effects which the project is likely to have on the environment;

    (iv) a non-technical summary of the information mentioned in paragraphs (i) to (iii) above."

Existing Environmental Statements

11.08 At our request, the Department provided us with actual Environmental Statements produced for seven schemes submitted since the Directive came into force. Four were concerned with village by-pass schemes in sensitive areas of the countryside, one with the upgrading of a motorway from dual two-lane to dual three-lane carriageways, and one with the construction of an outer by-pass around

---

[20] See Highways Act 1980, s105A.

a city. The seventh concerned the construction of a new route crossing an estuary involving a major bridge and was submitted in support of a Private Bill, the scheme being promoted in Parliament rather than through the ordinary Public Inquiry process. For most of the cases which went to Public Inquiry, the Inspector's Report and the Decision Letter of the Secretaries of State were also available to us. Some of the Environmental Statements had been prepared for the Department by consultants, others by the County Highway Authority as its agent.

11.09 The statutory requirements set out above are that the Environmental Statement should contain: (a) a description of the scheme and site; (b) a description of the measures proposed to mitigate any adverse environmental effect; (c) the data required to identify and assess the project's main environmental effect; and (d) a non-technical summary. It must also contain a statement of the main alternatives studied by the Department, and of the main reason for choosing the Preferred Scheme. The Departmental Standard HD 18/88 indicates that the Framework derived from the MEA will be the appropriate method for setting out the environmental data. The remainder of the material is to be set out in the text of the Environmental Statement.

11.10 With one exception (the estuarine crossing), the Environmental Statements which we have seen were surprisingly brief. The text, exclusive of the MEA Framework and accompanying drawings and figures, varied in length from 4 to 41 pages. One Environmental Statement failed entirely to mention the purpose of the scheme, which was (as the Inspector's Report showed) that of environmental improvement. The others contained brief but useful statements in the "Description of Scheme" section. One actually omitted the non-technical summary. The fullest treatment of environmental effects was contained in the Environmental Statement dealing with the largest scheme proposal, the estuarine crossing.

11.11 At one of the Public Inquiries an objector had argued that the Environmental Statement produced in that case did not fully comply with the requirements of the Directive. The grounds of the objection are not recorded in the Inspector's Report; but it was overruled by the Secretaries of State in their Decision and, so far as we are aware, that legal challenge was not taken on appeal to the High Court. We are not in any way concerned with the legal question as to whether the Government currently complies with the requirements of the Directive or the Highways Act of 1980. Historically, the Department has played a leading role in developing highway appraisal methods, and applying them to environmental impacts. Our concern is to recommend ways in which appraisal practice can be improved.

11.12 Experience of handling the requirements of the Directive is developing. In our view, the full opportunities offered by this form of presentation have not yet been grasped. The present format has been strongly influenced by the longer-established Framework, which in any given case is brought into the assessment process at a much earlier stage in the development of a scheme than the Environmental Statement itself. Indeed, it seems that a view has been taken that the Framework already contains the core of the information required by the Directive. If one adopts this approach (which may well be correct, legally) all that appears necessary, in order to comply with the requirements of the Directive in full, is to supplement the Environmental Statement with the rest of the required material not already contained within the Framework. Thus, in every case which we have examined, the Framework had been reproduced without any adaptation, including all the economic data produced by COBA, as well as the summary of the environmental effects, as the centrepiece of the Environmental Statement.

The Use of the Framework in Environmental Statements

11.13 We believe that the practice of including the Framework in the Environmental Statement should be abandoned. The introduction of material from the Framework has had the effect of both dividing and weakening the treatment of environmental effects. It has divided it by introducing data unconnected with the environment, and weakened it by confining the presentation of environmental data within the Framework's summary form only. It has, therefore, become unnecessarily difficult for a reader to understand the overall and particular effects which the scheme in question gives rise to. In this regard the Environmental Statements which we have seen were much more valuable where they dealt with matters outside the format of the Framework.

11.14 The Framework, as presently formulated, carries with it the danger of inducing a rather mechanical approach to environmental assessment. This can conflict with the essential thrust of environmental analysis, which is to identify and account for environmental effects, of every kind, to which specific proposals give rise. We accept that the organisation of relevant information on environmental effects in a standardised way is an important aid to public understanding of the differences between route options and to the process of selecting a preferred route; and that there is great value, in appropriate cases, in arranging data in tabular form as one method of presentation. We also recognise the advantages of the availability of a standard form of presentation. However, we believe that the Framework, by the constraints of its format, invites over-simplification, and an excessively rigid and inflexible approach.

11.15 For the same reasons, we do not believe that the environmental assessment, however it is presented, should be tied to the six Groups currently defined in MEA. To a significant extent these Groups (except the economic groups) are subsumed in the wider concerns which are contained in the statutory requirements of the Highways Act, and the subject-matter of some of the Groups will continue to play an important part in the assessment. We refer in particular to Group 4 (policies for conserving and enhancing an area) and Group 5 (policies for development and transport). In essence, however, we envisage the Environmental Assessment Reports and Statements of the future as being expressed in a greatly expanded and unconstrained form, more sensitively attuned to the special features of each proposal. Some more positive and constructive advice on this topic is contained in the remainder of this Chapter and the next.

Development of the Environmental Statement

11.16 We have considered the level of detail which is appropriate for consideration of environmental effects at different stages in the assessment process. Clearly, strategic assessment is, by definition, broad in grain. At early stages in the development of a scheme, it is also the case that the nature of any development proposal is not defined in detail, nor have detailed data been obtained about the site. Assessment at these stages will therefore be conditioned in terms of detail by these factors. However, for the stage of presenting alternative route-options at Public Consultation, we expect that all environmental effects of each option will have been identified and be included in the "Environmental Assessment Report - Route-Options".

11.17 The Environmental Statement will be prepared from the earlier environmental assessment and will relate specifically to the Preferred Scheme. It will be concerned with those effects which have been judged to be significant and, thus, not all effects considered at the Public Consultation stage will necessarily feature in the Environmental Statement. It is also the case that environmental effects which were either not considered prior to the Public Consultation stage or not considered to be significant, may feature as significant effects as a result of that process and will require to be included in the Environmental Statement.

11.18 As to the form of the Environmental Statement, we recommend that it be divided into sections and (in comparison with the present style) be expanded, as follows:

Section 1 **Description of the Site**
A fully detailed description of the site, and its main environmental features.

Section 2 **Description of the Scheme and its Environmental Effects**
A clear statement of the main objectives of the scheme, a full description of the scheme and a statement of its effects and the manner in which it relates to the principal environmental features listed above.

Section 3 **Mitigation of environmental effects**
Consistently with our recommendations contained in Chapter 12[21], this section will contain clear descriptions of measures to mitigate, avoid, offset or repair any environmental impacts which have been identified. Where this has been considered feasible and is part of the scheme proposals, measures to replace assets lost as a result of the scheme will also be described.

Section 4 **Data and methodology**
A full presentation of the environmental data. There is no reason why it should be presented in a summary form in this section of the Environmental Statement, because the purpose of the non-technical summary is to draw it together and present it in plain language at the end.

Section 5 **Non-technical summary**
The avoidance of technical jargon is especially important here.

Frameworks and Assessment Summary Reports

11.19 We wish to conclude this Chapter by relating now what we have said to the views expressed by SACTRA in the 1986 Report. That Report carried the more general recommendation that the Framework method of presentation should give way to a new format to be called the Assessment Summary Report. As is now apparent, our more specific consideration of environmental assessment has reinforced that conclusion in our minds. The Framework in the special technical sense in which that term has been used in this Report should now be superseded by the Assessment Summary Report for all purposes.

---

[21] See paragraphs 12.19 - 12.25.

11.20 The form of the Assessment Summary Report will itself need careful consideration, for it is here that all the different effects, separately and more fully analysed in the antecedent documents, will be drawn together and balanced out against each other. It is likely that the non-technical summary appearing at the end of the Environmental Statement will provide the text on environmental matters. It is also to be expected that a tabular arrangement of some or all of the data relevant to the scheme in question will be required, particularly if some environmental effects are monetised (see Part V of this Report).

11.21 It is however important to ensure that the Assessment Summary Report does not simply become the Framework by another name. The Department will no doubt wish to give guidance on the drawing-up of Assessment Summary Reports; but it should stress that the content of the final balance sheet must depend upon the particular circumstances of each scheme and warn against a mechanistic approach. Any suggestion that the Assessment Summary Report should be written to a mandatory format must also be avoided.

CHAPTER 12
THE MANUAL OF ENVIRONMENTAL APPRAISAL

12.01 The proposals contained in Chapters 10 and 11, if accepted, would have a considerable impact on the MEA. We have attempted to assess the merits of the existing Manual, both in association with our other recommendations and, more generally, in the light of the evidence we have received on the Manual itself.

12.02 In Chapter 6, we described the main structure of the MEA. It falls into three Parts: A, B and C.

Part A is concerned with the construction and preparation of Frameworks for trunk road appraisal. The purpose of the Framework is, as we have noted, a simplified presentation of results of other, more complex studies. It is intended to give a reliable summary of all the economic and environmental effects on the six interest Groups[22] of different scheme-options in order to facilitate a comparison between each option and Do-nothing or Do-minimum, and between the options themselves.

Part B is the most substantial part. It contains detailed technical guidance on the measurement and description of eleven specific environmental effects[23], which are then carried into the Framework.

Part C is supplemental to the other two parts. It is a select bibliography of further reading on both the formulation of frameworks as described in Part A and the environmental appraisal as discussed in Part B.

12.03 We have recommended, in Chapter 11, that the Framework method for the summary presentation of the results of both environmental and all other effects should be omitted from Environmental Statements, and be replaced for that and all other purposes by the Assessment Summary Report, as recommended by SACTRA in 1986. It follows that, in our view, the Manual should concentrate exclusively on environmental assessments. To distinguish it from its previous form we suggest that there should be a new manual named "The Environmental Assessment Manual".

12.04 The new manual should be a guide to the selection of the relevant environmental effects, to their description and quantification, to their assessment in terms of significance and to their amelioration should that be possible. It should give guidance as to the appropriate stages at which the work undertaken should be set in its policy context and assembled into a document. The checklists of environmental effects should indicate the scale over which and the time at which they are most likely to be critical. Whilst we recognise that it is Departmental practice to issue Departmental Standards and Technical Memoranda on matters of procedure and practice, very few of these are concerned with matters of the kind which we refer to above. On balance, however, we would prefer to see all relevant advice on these, and related matters, contained in one source document. The document could be presented in a form where sections could be replaced from time to time as advice developed.

---

[22] See paragraph 6.03.
[23] See paragraphs 6.07 - 6.09.

12.05 In its new role, the Manual will give technical guidance for the preparation of Environmental Assessment Reports and the Environmental Statement. This guidance should include advice on the use of specific expertise where the nature of the appraisal requires knowledge and experience beyond that which can be expected of the engineering team. As understanding of environmental problems, and the technology associated with them, advances it becomes more likely that other relevant expertise will need to be brought in. It is essential that this aspect of advice should be kept under review.

12.06 As to the technical rules for the appraisal of individual effects, they will form part of the new Manual, as is now the case, but will require to be expanded to cover matters not previously dealt with eg wider scale effects, such as may bear at regional or national level or the effects on natural resources (compare Figure 2). As an illustration of the scope of The Environmental Assessment Manual that we have in mind, Figure 4 sets out a possible structure for, and content of, the new format envisaged. We recommend that this be an important guide for the Department in producing the successor document to the Manual of Environmental Appraisal.

12.07 In the paragraphs which follow we offer some observations on matters related to the bringing forward of the new Manual which we consider to be of particular importance. These are concerned with the wide ranging nature of environmental assessment and the consequential discretion open to assessors; the policy effects of schemes; and the treatment of such matters as preservation and mitigation.

Additions to the Checklist of Impacts

12.08 As we note in paragraph 9.06, the assessment of the environment should ensure that every relevant environmental effect is fully described in appropriate terms, and that no effect is overlooked. Since every scheme design, and every location where such a scheme is to be constructed, creates a unique combination of environmental impacts, there is no pro forma which can cover all the possible effects in every case. Checklists are helpful but the responsibility for identifying all the particular effects associated with any particular scheme rests with the promoter of that scheme. This applies to highways with greater force than to other developments, because in this case the Department of Transport is both the developer and the body which grants permission.

12.09 We advocate a checklist of possible effects with guidance in the new manual as to how they could be measured and assessed and their significance judged. It is as important that the method of selection of the effects judged to be significant should be described as that the nature and scale of those effects should be demonstrated.

12.10 As the objectives of the scheme are likely to contain some sought-after environmental benefits, these should obviously be included together with any others deemed to support the scheme. It should be for those undertaking the work to justify which environmental effects should be the subject of appraisal, and which method of appraisal to adopt. The presumption must be that the selection will be rigorous and systematic. It should not be a matter of routinely following a particular list of effects, or that the depth of appraisal should be set regardless of the degree of significance.

FIGURE 4
Suggested contents list for the new Environmental Assessment Manual

PART 1	ENVIRONMENTAL ASSESSMENT
1.1	What is environmental assessment?
1.2	The objectives of environmental assessment
1.3	Strategic environmental assessment
1.4	Assessment Reports and the Environmental Statement
1.5	Assessing "relevance" and "significance"
PART 2	**UNDERSTANDING AN ENVIRONMENTAL ASSESSMENT**
2.1	The nature and scope of matters and effects to be considered
2.2	Climate and air quality
2.3	Effects on soil and water
2.4	Effects on flora and fauna
2.5	Landscape
2.6	Effects on people as individuals
2.7	Effects on communities
2.8	Effects on material assets and cultural heritage
2.9	Avoidance, reinstatement, mitigation and amelioration
2.10	Strategic and long - term effects
2.11	Interactions between the above factors
2.12	Time scales and discounting
PART 3	**EFFECTS ON POLICY**
3.1	Policies for conservation
3.2	Policies for Environmental Improvement
3.3	Development policies
3.4	Transport policies
3.5	The wider spectrum - environmental policies
PART 4	**TECHNIQUES OF ASSESSMENT**
4.1	Air pollution
4.2	Water pollution
4.3	Ecological impact
4.4	Visual impact
4.5	Land - take
4.6	Effects on open space
4.7	Effects on agriculture
4.8	Traffic noise
4.9	Personal stress and other effects on individuals
4.10	Disruption due to construction
4.11	View from the road
4.12	Community severance
4.13	Other effects on communities
4.14	Heritage and Conservation areas
4.15	Vibration

PART 5	PRESENTING THE RESULTS OF ASSESSMENTS
5.1	Organising the presentation of assessments and the Environmental Statement
5.2	The Environmental Assessment Report at Public Consultation Stage - assessing the options
5.3	Preparing desciptions: * the proposal * the corridor * the site * alternative development proposals considered * aspects of the environment likely to be affected * the effects on natural resources * forecasting methods * measures to eliminate, reduce or offset environmental effects * treatment of time scales
5.4	The non - technical summary
5.5	Other matters
PART 6	**GUIDE LIST OF POSSIBLE ENVIRONMENTAL EFFECTS**
	APPENDICES

12.11 It is evident that effects with an immediate impact which is local to a road line may be more readily anticipated, and it is on this that the present MEA has concentrated. As the appraisal widens to take account of a greater range of effects, a more sophisticated, hierarchical appraisal system will be required.

12.12 With regard to the preparation of a checklist within the new Environmental Assessment Manual we offer a number of observations. The 1986 SACTRA Report "Urban Road Appraisal" suggested two additions to the list: night-time noise, and the social and environmental effects of blight and development potential. In its Response, the Government accepted that recommendation, subject to the carrying out of further research into implementation. The fruits of that research have not yet been presented.

12.13 The 1986 Report also recommended the omission of View from the Road. The Government accepted this proposal only for schemes where effects do not differ significantly between options. However, for inter-urban schemes View from the Road must, in our opinion, remain an important factor. SACTRA in its 1986 Report also wished to exclude Driver Stress and Effects on Pedestrians and Cyclists. Neither seems to fit logically into the rest of MEA's list of eleven topics, the remainder of which are concerned with effects on the environment generally, rather than specific groups of travellers or other individuals.

12.14 Many other candidates for inclusion within the current MEA list have been proposed. From the various representations which we have received, we recommend that the following should be included in the checklist:

    a. **Land-take:** At present land-take appears in the economic assessment, in that the compensation payable for acquisition is part of the capital cost of the scheme. Currently, it also appears in the environmental assessment, but only insofar as it is relevant to any of the six Groups identified in the framework, notably 2, 3, 4 and 5. We believe that it should be considered as a specific topic in the appraisal, so that its full effects can be considered.

    b. **Loss of Open Space:** We believe that this too should be considered as a separate topic, for the same reasons as set out in a. above. Again, loss of open space only applies in the assessment at present, insofar as it is relevant to Group 3 (users of facilities) or 4 (policies for conserving and enhancing the area).

    c. **Water Pollution:** Road construction can have at least two effects on existing rivers, streams and ground-water regimes: run-off of storm water, and the permanent diversion of underground water system. These may affect quantity and quality of supply or an existing ecology and have some toxic effects. All these matters ought to be evaluated.

    d. **Vibration:** Some schemes, particularly in urban areas, may cause a local nuisance suffered as a result of vibration. In these circumstances, relief from traffic vibration is perceived as a considerable environmental advantage, and any change in the level of this type of nuisance, upwards or downwards, ought to be reported, difficult though it is to measure.

    e. **Personal Stress:** We do not believe that the present headings - community severance, driver stress and effects on pedestrians and cyclists - necessarily capture all the impacts of road schemes and traffic on human beings, individually or communally. Those existing topics should be amalgamated into a larger and more general category which deals with the human factor. It will then encompass all the numerous impacts on public health and well-being.

12.15 In addition to the inclusion of these matters in the checklist to be contained in the new Manual, we further recommend the redefinition of the grouping of the generic environmental impact headings as currently listed in Part B of the MEA. We find that the present arrangement is somewhat inflexible and it does not give confidence that areas of particular concern, such as effects on natural resources, socio-economic effects or strategic effects, are likely to be covered in the assessment process.

12.16 First, if the recommendations contained in Chapter 10 of this Report are accepted, and environmental appraisal is henceforth to be set more explicitly in a larger context of national environmental policy, the Manual must give fuller guidance on how this is to be carried into effect. At present, Groups 4 and 5 in the framework are intended to pick up the policy effects of schemes; but, as we commented in paragraph 6.08, the present format imposes some practical constraints on the manner in which that assessment is carried out. The policies themselves, of course, are developed by the Government, and set out in White Papers, Ministerial statements and other published material. They are not suitable for inclusion within the new manual. However, we would expect it to give guidance on the manner in which such policies are to be ascertained, and their general effect is to be assembled, so that the identification of routes and schemes, in so far as they significantly affect those policies, can be measured against them. Similarly, provision must be made for an appraisal of their effect upon any relevant policy adopted by local planning authorities.

12.17 In practice, this may be more difficult to implement than it seems, because we are concerned with an ever-shifting focus. We have already argued that, in the case of many local road schemes, the differences between options will often be too small to have any significant effect upon national or regional concerns, even though, at a local level, those same differences may be quite dramatic. It is at the very much earlier stage of the choice between different transportation strategies, and the identification of routes rather than schemes, that much of national environmental policy will come into play. We have also stressed the importance of the initial environmental appraisal which is carried out at the Route Identification Stage, and have called for a greater measure of codification of practice at that stage. In our view the new manual will be the proper medium for carrying out that recommendation.

12.18 Further, while we acknowledge[24] that, by the time individual schemes fall to be considered, both at the Scheme Identification Stage and subsequently, when options are presented for Public Consultation and Public Inquiry, many of the principal policy questions will have been settled, we reiterate that that will not hold true of all policies or for all schemes. Thus, the new manual must provide more fully than does the MEA for the explicit measurement and valuation of schemes by reference to such policies, so that if they are likely to affect them in some significant way that will not be overlooked.

---

[24] In paragraphs 10.11 and 11.01 - 11.03 above.

Preservation and Mitigation

12.19 The EC Directive, as incorporated in the Highways Act 1980[25], requires "...a description of the measures envisaged to prevent, reduce and, if possible, remedy significant adverse effects" on the environment[26]. We believe, however, that the execution of this requirement should be taken further than is the case in Departmental Standard HD 18/88. Specifically, we believe that in HD 18/88 the sense of paragraph 2.6.1 should be widened to include the possibilities of preventing or offsetting a significant environmental effect as well as simply reducing it. In making this proposal, we have in mind such possibilities as the relocation or the re-creation of environmental assets which may be significantly affected and are of a kind amenable to measures of this sort. We consider that paragraph 2.6.2 should also be amended to reflect this line of thinking.

12.20 The former Nature Conservancy Council, in its evidence to us, has advocated that where an SSSI is affected by a road proposal, there should be an assumption either that remedial works will be undertaken or that there will be an equivalent reinstatement. We believe that, where an asset of environmental quality is threatened, then a formal sequence of reviews should be carried out by the environmental assessor. These reviews can be expressed as three successive questions. These are, first: "Can this damage to quality be prevented?". We expect that an answer to the question will only usually be possible after physical and environmental studies with this objective in mind have been conducted. We believe that a summary of such studies should be included in the Environmental Assessment Report and, if avoidance is incorporated into the scheme, be incorporated in the Environmental Statement.

12.21 The second question is "If the damage cannot be avoided, can it be reduced?". By this we mean whether there is some compensating action or actions which ameliorate the extent, or the intensity of the damage, even if damage cannot be totally avoided. Again we expect this matter to be incorporated in the manner described in the previous paragraph.

12.22 The third question is "If the damage cannot be avoided or reduced, can it be offset?". By this we mean whether something else could be done which creates a benefit to set against the damage. For example, in certain cases it may be possible to re-instate the asset or benefit which will be lost at some nearby location, or to create an environmental quality of a similar kind where quality has been lost because of the scheme. In an appropriate case, measures of this kind could properly be regarded as a form of "remedy" and may be the only type available. As before, we expect these matters to be reported upon.

12.23 It follows that every assessment, where this is relevant, ought to state not merely the remedial measures necessitated by a scheme, but also the most cost-effective way of carrying them out. When the remedial works are in fact going to be performed, the costs are already brought into the reckoning as part of the costs of the scheme. The same is true in cases where a road has an adverse impact on land, and statutory compensation is payable, such as for severance or injurious affection, or double glazing where adjoining owners are likely to be troubled by noise.

---

[25] By SI 1988 No 1241.
[26] Annex III, paragraph 5.

12.24 In other cases of environmental damage, where expenditure is not going to be incurred or compensation paid, estimates of remedial or replacement costs can at the very least serve the useful purpose of obliging decision-makers to focus on the scope of the physical damage involved, and consciously to balance the value of the asset in question against that cost, in order to form a view of its relative importance.

12.25 Obviously there are limitations to the circumstances in which this type of assessment can be carried out. Sites having historic or cultural associations, or large scale landscapes, may be susceptible to protective measures such as screening, but not much more; and the method may be inapplicable altogether in the case of air pollution. But in general it can be applied to a wide range of diverse physical impacts.

12.26 The existing Manual of Environmental Appraisal, in its present form, has the great virtue of being familiar to many practitioners and indeed many that gave evidence to us did not consider that it needed significant alteration. Its existence has been of very great benefit for the Department over the years since it was conceived and developed. In a number of respects it gave an environmental dimension to decision-making about large scale investment in Britain which was highly innovative; and it was in service long before the EC Directive 85/337 was incorporated into our legislation. It has therefore been only as a result of the most careful discussion that we have come to recommend the character and extent of change which we have outlined in this Chapter. We make these recommendations, however, in the light of the increasing expectation on the part of the public of the highest standards of environmental stewardship over many areas not previously recognised.

PART V

THE MONETARY VALUATION OF ENVIRONMENTAL EFFECTS

In this Part of the Report we address the specific questions raised in our terms of reference: whether a greater degree of valuation of environmental effects is desirable, what would be the appropriate scope and application of such valuation, and what methods would be suitable for deriving monetary values.

CHAPTER 13
VALUING THE ENVIRONMENT: PRINCIPLES

13.01 This part of our terms of reference has required us to reconsider the view clearly expressed by ACTRA and SACTRA in their earlier Reports[27] that environmental benefits or costs should not be evaluated in money terms, but that their assessment should lie exclusively within the realm of professional judgement. That view will have coincided with the instinctive reaction of many people to the same question. What monetary value, it may be asked, can you put on an Area of Outstanding Natural Beauty or an historic monument, or a rare species? Even to make the suggestion may engender some suspicion - a lurking fear that once it has been shown that, for example, part of our national heritage is "worth" £x, cost-benefit analysis will be able to prove that it should not stand in the way of an important road scheme, and it will be destroyed.

13.02 Indeed, some would go even further, and argue that cost-benefit analysis should play no part in the appraisal process at all, not even to the extent achieved in COBA. In its place we should formulate a set of objectives which our transportation system generally, and roads in particular should set out to achieve - economic, environmental and all other. Policies and schemes should then be appraised according to the extent to which they achieved or failed to achieve the various objectives which the Government set. Inevitably, options will exist, and there will have to be trade-offs between different objectives. It would then be a matter for political judgement to decide whether the achievement or near-achievement of some objectives should be accorded a greater degree of importance than the achievement or near-achievement of others.

13.03 The SACTRA 1986 Report, with its emphasis upon the need to appraise schemes in accordance with their ability to contribute to objectives, or to solve stated problems, might appear to lend some support to this argument. It is indeed important to examine the performance of schemes and policies in relation to stated objectives, but this does not imply, and it was not implied in the 1986 Report, that appraisal methods should be informal or merely descriptive.

13.04 The choice between formal and informal methods depends upon the context in which the decision is made. In the roads sector, appraisal is used throughout the planning process. Every year, numerous decisions on scheme design and routeing are taken. The few decisions taken to proceed or not with a scheme are at the top of a pyramid. Appraisal is used at decentralised levels throughout the Department, and by its consultants. We concur with the earlier Reports of ACTRA and SACTRA that a formal method of appraisal is required to furnish decision-makers with the factual and technical material upon which their decision is made, and to enable them to give any one effect of a proposed scheme or policy - on cost, time, safety, environment - its due weight against all others,

---

[27] See Chapter 5 paragraphs 5.04 - 5.05 and 5.07 - 5.10 and Chapter 7 paragraph 7.07 above.

whether they are of a similar or a different character. Such a method is not a substitute for decision-making. It is a tool in the process. Without it, it is difficult to see how decisions of the type described in paragraph 13.02 could be made in a rational, explicit and consistent way, and in a manner which would be generally accepted and understood.

13.05   One of the special characteristics of the appraisal of road schemes is that it involves forecasting the impacts of schemes in numerous dimensions. The decision-maker has to be presented with, in effect, a balance sheet comprising a large number of very disparate elements. It follows that if some common unit could be found to measure impacts of a seemingly different character, the difficult task of balancing them out might perhaps be simplified, and a greater consistency achieved between separate decisions affecting similar subject-matter.

13.06   There are plenty of techniques available for combining different effects: converting qualitative assessments into common categories such as "negligible", "slight", "moderate", and "severe"; drawing up a matrix and inserting symbols such as ticks, blobs, and crosses to convey the same message; points-scoring, attaching to each impact a weighted score reflecting the relative importance being placed on the various impacts, which is then summed overall; and colour-coding. In their various ways all these techniques stand as proxies for a series of possibly complicated judgements on individual issues, with the intention that they can be more easily traded off against each other. In this regard, we were impressed by the efforts made by the Department and its Consultants to present their findings on the London Assessment Studies to the public in an accessible form.

13.07   Cost-benefit analysis uses money values as the relative weights to be placed on different attributes or effects. The use of money as the standard or numeraire in appraisals has a variety of attractions which are additional to those of the techniques described above. Evaluation in money terms provides a basis for comparative assessment of value for money across different transport sectors. There is also now a tradition of using cost-benefit analysis in the roads sector, where some effects such as construction and operating costs are naturally expressed in money terms. Cost-benefit analysis leads to greater efficiency in the use of economic resources, by enabling decision-makers to see where such resources might be wasted, and where they could be best deployed.

13.08   Monetary values also have their limitations. They normally imply making simplifying assumptions about the homogeneity of the attributes to which values are attached. Standard values are often employed. A degree of averaging and coarse categorisation is required in order to make the appraisal manageable. Some of our witnesses have suggested that the use of monetary values cannot be extended comfortably to the environmental dimensions of roads because of their innate heterogeneity. Thus the value of a given noise reduction might, it is said, be dependent on the circumstances, and not be capable of approximation using some standard table. Again, there are some environmental factors which cannot sensibly be valued in monetary terms at all. For some, this would be the end of the argument, for it is then suggested that once it is conceded that there are some environmental factors which cannot be valued in monetary terms, it would be wrong to attach values to any of them, it no longer being possible to find the "common currency" which was the object of the valuation exercise in the first place.

13.09   There is force in these criticisms, which were reflected in much of the evidence which we received.

A further argument against extending valuation in this direction is that effects which are valued are habitually given priority over those which are not. Thus, despite the advances made following the ACTRA Report, environmental effects are still thought to receive less weight than (say) those affecting road users. Correspondingly, there is now concern that if some environmental effects were to be valued, they would be of greater influence than those which are not. In this regard it is to be expected that environmental effects which are immediately perceived by people to affect them in their personal lives, such as noise and similar effects on amenity, would be valued more readily than those bearing upon, for example, the natural environment.

13.10 At present, as we have shown, the appraisal of road schemes already contains a mixture of methods; and in that respect the present practice is unlikely to undergo any fundamental change. For the reasons set out in paragraph 13.07, we believe that there are over-riding advantages in applying reliable valuations to environmental effects wherever they can be made; but we also wish to emphasise that a particular responsibility is placed on those carrying out the appraisal to ensure that no bias is introduced between those effects which are valued and those which are not.

13.11 We have had extensive discussions about the issues of theory and principle involved in the valuation of social and environmental impacts, their aggregation, and the implications of social cost-benefit analysis. It is not necessary to extend the text of this Report by setting them out here. The interested reader is referred to Pearce, Markandya and Barbier, among others[28]. The conclusions we have reached are as follows:

(1) There is no legitimate objection of principle to the use of monetary values for evaluating as many of the environmental effects of road schemes as lend themselves to that technique, even if others cannot be so valued.

(2) There are great advantages to be gained from extending monetary valuation as far as it is reasonable to do so.

(3) However, any methods suggested for such valuation must satisfy the criteria laid down by us in paragraph 9.06 of this Report.

(4) There is a danger, which must be recognised, that as more environmental effects are introduced into cost-benefit analysis, a bias in favour of those effects may result. Those which are not valued must still receive their due importance in the appraisal.

---

[28] See paragraph 1.08, footnote.

CHAPTER 14
REVIEW OF EXISTING VALUATION TECHNIQUES

14.01   As we have seen, environmental impacts are disparate in nature, and it is clear that no single analytical method can be used as a basis for valuing all such impacts. Diverse types of impact imply a variety of valuation techniques. There is also a scale of relative difficulty in valuing different types of effect. At different points along that scale different techniques are called for.

Environmental Effects Which Cannot Sensibly be Valued in Money Terms

14.02   We stated in paragraph 13.08 that there are some environmental effects which cannot sensibly be valued or expressed in economic terms, and it would be contrary to common sense even to attempt to do so. For example, if global change, leading to damage to the ozone layer or to the greenhouse effect, were to take place on such a scale as to jeopardise the survival of human society, it would be pointless to value the difference between "Do-something" and "Do-minimum" in order to assess whether the survival of the species warranted the cost of protective action.

14.03   Further, in practical terms, there is little to be gained from ascribing monetary values to some unique or sacrosanct environmental assets. The loss of such assets will always be a matter for overt and explicit political judgement. Those judgements are unlikely to be better informed by monetary valuation.

14.04   A third example (which may overlap with the two mentioned above) is that of irreversible cumulative impacts which are felt by future generations. Here the argument is different. It would seem wrong that the values of the current generation should be imposed on those in the future who may not share them.

14.05   We conclude therefore, that there is a class of environmental impacts - potentially catastrophic changes, losses of unique or sacrosanct assets, and long-term impacts on future generations - for which monetary valuation techniques are unlikely to be helpful to decision-makers. But we must avoid the paradox whereby important impacts which cannot be valued in money might somehow be undervalued in the appraisal.

Actual Costs

14.06   At the other end of the scale it can be demonstrated that some environmental effects, like many other effects, have an immediate monetary consequence which results in the incurring of actual expenditure. Indeed, some of the costs already included in the appraisal fall into this category. As we pointed out in paragraph 12.23, whenever, in building a road scheme, it is necessary to carry out mitigating measures such as the planting of trees or the construction of other landscaping features, or to pay statutory compensation to adjoining owners for double-glazing to protect them from noise, all such expenditure is brought into the reckoning for the purposes of cost-benefit analysis.

14.07 The point about these costs is that they are all incurred either by the agency promoting the scheme - the Department of Transport - or some other branch of government. Logically, similar heads of cost should also be included whenever (as frequently happens) they are borne by a third party, provided that they can be reliably ascertained. If, for example, water run-off from a road results in downstream pollution which a private individual - a farmer who uses the supply for watering cattle or an industrialist who uses it for a factory - incurs expense in remedying, proper account should be taken of this factor too.

14.08 In paragraphs 12.19 to 12.25 we have already emphasised the importance of including within every assessment a full analysis of all measures available to protect environmental assets, or to mitigate the adverse effects which a scheme might have upon them. The Environmental Statement itself must address these questions, as a matter of statutory obligation. The carrying-out of that work should, as we have previously said, comprehend the investigation into costs which we are discussing here.

14.09 No breach of principle is involved in expanding analysis in this direction because, as we have also seen, third party costs and benefits within the value of time, the value of accidents, and vehicle operating costs are already taken into account. The only change is one of practice: how far is it possible to go, in practical terms, in identifying these equally direct but less easily detectable economic effects? The question is simply a matter of confidence in the analysis of impacts, not an issue of principle.

14.10 At this point we draw attention to two special cases. In their different ways they each constitute a modification or extension of this simple principle.

Special Case (1): The Cost of Land-Take

14.11 The capital costs of the scheme which are encompassed within the COBA program include, in part, the cost of acquiring the land required (if necessary, compulsorily) from those who own it. The amounts entered into the program are the actual amounts to be paid to the dispossessed owners. Those amounts are calculated in accordance with the statutory rules laid down in the Land Compensation Act 1961 and other Acts, and are normally based upon open market value - the price which the land could reasonably be expected to fetch if sold on the open market by a willing seller to a willing buyer. That price will reflect (1) the land's existing use value and (2) any potential which it might have, in the open market, for development for other purposes (its alternative use value). It cannot, however, reflect any enhanced value realised as a result of the actual scheme for which it is being acquired. It has to be valued, as it is sometimes put, "in a non-scheme world".

14.12 Much of the land acquired for road-building has a low existing use value, because it is in the countryside, or is open space. For these same reasons it may have a high environmental value. It is also unlikely, under our system of development control, to have much development potential. Very often it has none at all. The result is that this element of the capital cost of the scheme is likely to be quite low; and the imperatives of the search for value for money may encourage a cost-conscious Government Department to prefer to take such land, rather than developed land, where the cost of acquisition is likely to be greater. It may, therefore, be said that there is a paradox here: the very system of development control which is designed to protect land which, in environmental terms, is of the highest value, causes it to show the best value for money and to be the most attractive, from a cost-benefit point of view, for the purposes of road-building. The "social" value of the land, it may be said, does not in any way correspond with the open market value which is attributed to it for the purpose of compensation, and is used in Route and Scheme Identification Studies and thereafter.

14.13 The paradox could be resolved if it were possible to find a convincing substitute social value to be used in cost-benefit analysis in place of (and necessarily in excess of) its value under the rules for assessing statutory compensation. The argument runs on the following lines. The restrictions on development imposed upon the land in question by planning controls reflect the importance to society of reserving the land for certain uses only, for example agricultural use or use as amenity land within the Green Belt. For the benefit of that restriction society has been willing to pay a price: the sacrifice of the opportunity to divert the land to any other use, that is, the sacrifice of its development potential. The acquisition of the land for road-building now frustrates the very purpose for which the land was being reserved. Therefore, for social cost-benefit analysis purposes, its value should equate to the price which society was previously willing to pay for the benefit, now removed, of the earlier restriction.

14.14 The argument has great attractions, but raises some formidable problems in terms of practical implementation. We can summarise them here:

    a. It is unclear what land should be valued for these purposes. It could be the strips of land actually acquired, the whole of the parcel of land of which they form part, or perhaps an even wider area.

    b. The alternative development value of that land will depend on the alternative uses to which, in the open market, it could be put, and its relationship, in that market, with other land competing for the same uses.

    c. For these purposes it will be important for the valuer to know what planning regime is assumed to apply to other potentially competing land. If existing planning restrictions remain in full force on all other land, the acquired land will enjoy an artificial scarcity value. Valuers might find the alternative assumption (that there are no planning restrictions anywhere) impossible to implement.

    d. Even if such an alternative value could be found, it may still be doubted whether it represents the true social value of what is lost.

14.15 We conclude therefore that there are circumstances where the expected financial cost of land for new roads is likely to be an underestimate of its social value, which ought to be the relevant value for cost-benefit analysis. We see practical difficulties in arriving at an appropriate basis for social valuation. However, we are clear that current practice is unsatisfactory and in need of review. In the meantime, this is a weak point in COBA which may significantly affect the ranking of options. It needs to be borne in mind in any decision-making which relies upon it.

Special Case (2): Mitigation which is not Carried Out

14.16 We also recommended (in paragraph 12.24) that the costs of conservation, mitigation, and clean-up should in any case be explicitly stated in the assessment, even when they are not going to be incurred in fact. This may occur because the agency on whom the cost would fall is not compelled to meet it, and cannot or chooses not to do so. In such a case, the agency has decided that the environmental damage is less important than the cost of preventing, or remedying it. In a sense, therefore, an implicit value has been put on the asset in question: the estimated full mitigation cost represents an upper value on the damage in question, the lower bound being zero.

14.17 We therefore conclude that even where a decision is taken, or where it is very likely, at the end of the appraisal process, that a decision will be taken, that environmental damage is not going to be remedied, the cost of remedy will nevertheless be relevant to decision-making. It should not be reported uncritically, but should be reported.

Values Revealed by Behaviour

14.18 There is another range of effects which attract real money values. These effects are confined to relatively well-defined groups of individuals, who perceive that they are being affected by some types of environmental conditions, and reflect those conditions in money which they are prepared to pay in the real market-place. Visual amenity and pleasant and peaceful surroundings are assets which command a price. People will pay a higher price for houses which enjoy these benefits than for those which do not. People at work too have become more discriminating about their physical surroundings. Commercial property in a pleasant location will be more attractive to staff and more impressive to customers, and will attract a higher rent. The commercial value of holiday and leisure sites is influenced in part by environmental amenity.

14.19 These phenomena are well-recognised by professional valuers, particularly in the residential property market. In the context of trunk road schemes, District Valuers (officials of the Board of Inland Revenue who perform a wide range of property valuation tasks in the public domain) have acquired a special knowledge and experience. Their role is best known in negotiating claims for compensation for land which is being compulsorily acquired; but in addition to that they have to deal with many claims which are linked to environmental changes brought about by road construction. Typically, they involve the evaluation of effects causing some degree of local nuisance - noise, vibration, smell, fumes, smoke and lighting. In arriving at appropriate values they are guided by a substantial data base of property prices ("comparables") and trends, to which they apply their professional judgement, aided by local and more general professional knowledge. Their counterparts in private practice, with whom they are regularly in negotiation, exercise the same skills.

14.20 This body of expertise is, in our view, at present under-used in the process of assessment. A reliable professional estimate of the effect of environmental change on property values must, to put it at its lowest, afford some guide to the relative importance which people attach to the environmental factors which are perceivable by them and directly affect them. If the properties are not of an unusual or special nature, then the reduction or increase in market value will be closely related to the loss or gain in welfare as it is perceived by the people who live there, and this information must, at the very least, assist in the process of making choices between options, to the extent that welfare considerations are relevant to them.

14.21 A considerable amount of academic research has been carried out in a number of countries into the same issue, using more formal statistical analysis of property values, rather than professional judgement. Again, the aim of the research has been to discover the extent to which consumers are sensitive to environmental factors and how their attitudes are reflected in the prices they are willing to pay.

14.22 The weight of the evidence we have received from this source seems to suggest that there is a discernible statistical relationship between house prices and some highly localised environmental factors such as noise or air pollution. A University of Salford study, for example, indicates that the reduction in property values within the noise contour of the Manchester International Airport might be as much as 6% which, as a proportion of the total value of a house, might be thought to be high. About 15 studies - mainly in the USA - of the relationship between aircraft noise and house prices are quoted in a review commissioned by the Transport and Road Research Laboratory[29].

14.23 These techniques have some limitations and weaknesses.

14.24 First, the price of a house represents a payment for a mixture of physical characteristics, location, access to job opportunities and other attributes, in addition to environmental factors. The academic research shows how difficult it may sometimes be to segregate these different elements in a manner which is statistically convincing, or likely to command general acceptance. Professional valuers might feel able to express a greater degree of confidence in their estimates but the opinions of valuers on issues such as these may in difficult cases vary considerably.

14.25 Secondly, it will often be difficult to judge the extent to which those price differences reflect the real importance of the environmental damage in question. If the effect is highly localised, such as noise, the relationship might be close. By contrast, the destruction of an important site may have far-reaching significance beyond its immediate neighbourhood. In such a case shifts in the value of local properties would be of little or no significance in the understanding of the impact taken as a whole. The weight to be attached to such evidence would be a matter of very great controversy.

14.26 Thirdly, even in the case of localised effects the application of the principle is necessarily limited. It is difficult if not impossible to operate where the housing market consists largely of rented property, especially public housing; and it cannot provide a value for the environmental factor in question for persons who are not also consumers in the housing market, such as non-residents or non-contributing members of the family.

---

[29] "Environmental Appraisal: a Review of Monetary Valuation and Other Techniques", report to the Transport and Road Research Laboratory by Rendel Planning and the University of East Anglia, 1990, unpublished. A summary version of this Report will appear in the Transport and Road Research Laboratory's Contractors Report Series in due course.

14.27 Another market-based approach is to attempt to interpret the observed behaviour of users of facilities such as historic buildings, areas of special landscape quality, or parks. It is known as the "travel cost" method. It attempts to gauge the value which people attach to facilities of this kind by assessing the level of their use, and the amount people are prepared to spend to get to them. It proceeds on the assumption, already built into valuation systems such as COBA, that the "cost" of a journey is the sum of the time and money spent on it. It then follows that anybody visiting the site must value it at not less than the cost of getting there. Not surprisingly, it is usually observed that there is a decline in use of such a facility with the distance from it. The rate of decline gives information about how much people are prepared to pay to use the facility. This can then be used to estimate the value of converting the site to another use, for example a highway or, if it could be shown that the demand curve would shift in response to changes, the value of the environmental gain or loss resulting from the changes.

14.28 There are recognised limitations on the application of this technique. First, the most important sites are rarely lost by outright conversion to highway use. Where they are only partially affected it is unlikely that there will be sufficiently close correlation between the proposed change and a shift in the demand curve for any confidence to be placed in the result. Secondly, the method is based on the assumption that the journey is purely a cost, and of no value in itself. No distinction is made between multi-purpose trips, and those undertaken exclusively for the purpose of visiting the site.

14.29 There are other problems too. The method accounts for use value only. It is inappropriate for sites which are rarely visited. It does not deal with "option value" (the value of preserving the option to visit in the future) or "bequest value" (the value to future generations). Nor can it accommodate changes in demand over time for one site in relation to other sites or areas of the country. At best it might in a special and well-defined local situation give decision-makers some rough indication of the importance which the public at large attaches to one site in comparison with others, but we doubt whether the results obtained by this technique, in its present state of development, could be converted into economic value with any degree of confidence.

Individual Values Expressed by Statements of Preference

14.30 Another way of finding out what values individuals attribute to different effects is simply to ask them. Provided that the question is asked in a way that is comprehensible, relates directly to an understood experience, and safeguards against the possibility of fraudulent or biassed answers, there are advantages. The topic of concern can be addressed directly without all the complicating factors of interpreting choices revealed indirectly by behaviour. The main disadvantage is that, not being proved by action, the answer may not be accurate.

14.31 The impacts that can be best addressed by this method are, like revealed preferences, those perceived at an individual level rather than a community level. Within the last decade, much of the work on travellers' preferences, including values of time and safety, has been of this genre. The values of time were regarded as important contributory evidence to overall cost-benefit analysis. Work on environmental issues is in its infancy. There are two methods that are in use in determining hypothetical values, the "contingent valuation" and "stated preference" methods.

14.32 The "contingent valuation" method asks respondents to state the maximum sum they are willing to pay to prevent the destruction of an environmental asset or for environmental improvements such as cleaner air or water. Alternatively, they may be asked to state the minimum compensation which they would be prepared to accept for its loss. The great advantage of this method is that the environmental good in question can be singled out for attention in a way which revealed preference finds difficult.

14.33 The method possesses other advantages too. It can (but with greater difficulty) be applied to a range of environmental factors beyond the somewhat narrow categories which are linked to user-values, such as property prices or travel cost. The method has been used to value a wide range of environmental goods in the USA including water quality, wildlife, visibility and air quality, recreational activity and forest quality. In the UK, the method has been applied to amenity assets such as beaches and also to important ecological assets (an SSSI and a nature reserve). It is not limited geographically, at least in theory, to the area in which the respondents live or have an immediate concern, and it can tackle the "option" and "bequest" values referred to above, which are beyond the reach of revealed valuation techniques. If it is to be extended in these respects, however, considerable care is called for in establishing that respondents are sufficiently familiar with the subject-matter to give unbiased and well-informed answers.

14.34 This question of bias constitutes one of the more serious problems in its practical application. First it has to be decided whether the right question is willingness to pay (for environmental conservation or improvement) or willingness to accept compensation (for detriment or loss). Sometimes quite different values are obtained, depending upon the option offered. Secondly, if willingness to pay is the criterion chosen, it is necessary to specify the means by which payment is made, whether it be through national or local taxation, the price of relevant goods or services (for example a toll tunnel) or some other means. Again, responses may be influenced by these variables. There is also the question of deliberate bias: respondents may state sums of money which are much greater than those which would in truth satisfy them, in the knowledge that no real money is on offer, no real charge is being threatened, and their answers may influence policy in a desired direction. However the Report to the Transport and Road Research Laboratory[30] states that bias in empirical contingent valuation studies is less common than theorists would predict. We think that further studies of this kind can provide useful background information for decision-makers on people's preferences, to be weighed in with other expert evidence on the importance of particular assets.

14.35 An alternative method is the "stated preference" method which is aimed at eliciting preferences by offering multiple combinations of goods or attributes which are rated or ranked by respondents. The most well established applications have included offering different combinations of time and money in the context of a hypothetical mode or route choice, and different combinations of comfort, amenity and convenience in the context of station or train design. The method has also been used in a wide variety of other speculative applications where real world applications are difficult to observe, for example the impact of opening a new facility which does not yet exist, or the provision of a new information system.

---

[30] See paragraph 14.22 above.

14.36   In this method respondents are never asked directly "how much do you value such-and-such?". They are presented with a number of different situations, and asked what they would choose. A well-designed investigation will tailor the options offered, and the varying levels of the attributes described, to correspond with the individual's actual circumstances in order to make the choice as realistic as possible. Because the levels offered are controlled by the experimental design, it is possible systematically to ensure that one variable is altered while others stay constant, so that some of the complicating circumstances of the real world are avoided. Logic checks can be built in so that the respondent (or the analyst) can be warned if the choices expressed are inconsistent or self-defeating.

14.37   The result of such an experiment is a series of decisions, predicted by the individual rather than actually made, corresponding with well-behaved and carefully structured alternatives in contrast with the alternatives usually on offer in the real world. To the extent that these decisions are to correspond with those that the individual would make, they can be interpreted using the same analytical techniques as those which would be applied to real choices, and the relative weights of the different factors can be estimated with very much tighter levels of statistical precision than is typically available from real choices, hence smaller sample sizes and cheaper research. It is difficult or impossible for the individual deliberately to defraud the system (it is just too complex to work out what answers to give to have a desired effect) and those with experience of operating such systems typically report that respondents make a genuine and serious effort to give well-judged answers.

14.38   The method is currently well-thought of by professionals, and lends itself well to valuation of a wide range of environmental amenities which are enjoyed by individuals, and to assessment of the losses they feel they would suffer by certain kinds of deterioration. There must always be a residual caution in using the method that its answers may not be what people would actually do, and it should be a matter of routine that in every case where going back to check is possible, this should be done.

14.39   Thus, there are some practical and fundamental problems with both of these techniques. We do however believe that the results of properly conducted surveys into contingent valuations and stated preferences can, in the long run be of value to decision-makers. The practical problems which we have described are well-recognised. It is likely that they will be surmounted over time and that further research will produce acceptable improvements in techniques.

Equity and Distribution

14.40   In both the revealed and stated preference techniques of valuation there is the general issue of the treatment of equity and distribution. The amount people are willing to pay is dependent to an extent on their means. This raises the question of how values derived from studies of house price variations or people's stated preferences are to be used in the appraisal process. Are they to be derived locally and employed on a case by case basis to value perceptions of changes in benefit and loss at a local level? Or are the results of a series of studies to be consolidated so as to provide a set of standard appraisal values to be applied across localities? There is difficulty here in that current procedures contain a mixture of approaches. For example, property is valued on the basis of its market value, whereas non-working travel time and accident benefits are valued using standard values which may differ from the actual willingness to pay of the particular group receiving the benefit. We think that there are arguments for using standard values for impacts of general application such as noise reduction which are as strong as they are for time and accident savings. If, however, there is to be a development from the case by case approach to standardisation, a substantial research programme will be required.

Values which derive from Policy Constraints and Targets

14.41   There are many environmental impacts in which the summation of the individual preferences of a specific group of users, no matter how accurately estimated, is not a valid measure of social worth. This especially relates to matters which are the subject of national, macroeconomic or global policy decisions, and which derive from the inadequacies of market processes, for example where there are hidden costs.

14.42   In such cases, there has been a tradition of calculating "shadow costs" which are not equal to those actually charged in a market, but are estimated more precisely to correspond with the real long-term resource costs of the economy. Areas in which this technique has been adopted include investment appraisal in under-developed countries, where there are very high levels of unemployment and the use of market wage-rates for labour costs would artificially overstate the real opportunity cost of carrying out an investment; and in foreign currency transactions, where premium notional prices have been assumed at a time when balance of payments problems are dominant. The more fully market prices actually incorporate full resource costs, the less necessary it is to make such calculations. This approach may also be applied in cases where over-riding policy objectives are concerned, and where the concept of a "trade-off" seems inappropriate. We consider the important case of sustainability.

14.43   Sustainable development is defined in "This Common Inheritance" (paragraph 1.14) as "not sacrificing tomorrow's prospects for a largely illusory gain today". The White Paper points out that ".... we do not have a freehold on our world, but only a full repairing lease. We have a moral duty to look after our planet and to hand it on in good order to future generations". It is clear that the process of defining detailed policies which will implement this moral principle is still continuing. But some indications are given in the White Paper. One example is "the demanding target, if other countries take similar action, of returning CO_2 to 1990 levels by 2005" (Summary of Government Action paragraph S.16).

14.44   If such a target were to be accepted, the next stage would be to consider how it might fit into the framework of economic appraisal. The easiest case is when the target would be achieved, in part or in full, by setting appropriate price levels. There are strong arguments for this to be done, as set out in the White Paper:

> "The aim of a market based approach to the environment is, therefore, to give consumers and industries clear signals about the costs of using environmental resources" (Annex A paragraph A.2).

> "There is a good case for charges which show polluters and others the costs which their use of the environment imposes on the community as a whole ..... a pollution charge is no more than a price for using scare resources" (Annex A paragraph A.17).

With this approach, the Government could then decide to set a price for petrol that would be designed to ensure that transport contributes its due share (still to be determined) to the overall target.

14.45 In that case, the implication for the appraisal of road schemes is clear but it requires a change from current practice in one important respect. At present, it is assumed that the taxation element of petrol costs (and indeed other taxes) is not a resource cost to be included in the social cost-benefit appraisal, but a "transfer payment", in which losses of economic utility by the payers are exactly balanced by equivalent gains from the eventual recipients of tax spending. The taxation is therefore "netted out" from the calculations. However, a pollution charge justified as a price for scarce resources has to be treated as a real resource cost. Any cost element, levied through the tax system but justified in terms of the environmental resources consumed should be incorporated in the evaluation.

14.46 Therefore, the Government policy setting an environmental target to be reached, and using the market to achieve it, must result in a corresponding entry into the cost-benefit calculations. In the example discussed, this would have the effect of increasing the net value of policies or schemes which reduced petrol consumption, and reducing the value of schemes or polices which increased petrol consumption, to a level designed to achieve the target. The price would then be set by considerations of what is known about short and long term price-elasticities for petrol.

14.47 However, we recognise that there may be political or administrative or practical difficulties in the way of using the price mechanism in this fashion. Alternatively, there may be other, more effective, ways of achieving the same target. In that case, the actual price charged for petrol would underestimate its resource value, and it would be appropriate to calculate a "shadow price". A "shadow price" represents the price that would be charged if the correct market values were being charged to the consumer. Then the same calculations can be carried out, to assess what the market price ought to be to give the level of consumption defined by Government policy to be consistent with its over-riding objectives of sustainability. That value can be entered into the cost-benefit calculations even though it does not correspond with the actual market price (in the same way as some other values currently used in COBA differ from actual market prices).

14.48 The same principle can be extended to the various other instruments available to achieve Government targets and constraints. In the White Paper, there is discussion of administrative controls, industry levels, charges and producers, tradeable quotas, subsidies, and changes in the base of legal liability and private compensation. In each case, there is a corresponding resource cost, explicit or implicit, which more closely represents the value of environmental damage than the costs used at present.

14.49 The advantage of this approach is that any binding constraint or established target may be represented as the economic cost which would bring it about (which is also to say that setting such targets may also cause a loss of other benefits). Its legitimacy does depend, however, on the assumption that the constraint or target is a correct one. If Governments set policies for the environment that are, in some sense, wrong, then so will be the corresponding economic values implied by those polices.

CHAPTER 15
THE WAY FORWARD

15.01 In response to our terms of reference, we have explained in Chapter 13 why a greater degree of valuation of environmental costs and benefits is, in principle, desirable; and in Chapter 14 we have reviewed, in outline, the various methods for deriving monetary values which are currently regarded as suitable for general use in environmental impact assessment. Whether any of these techniques is sufficiently advanced to be incorporated immediately into the appraisal of road schemes has been the subject of detailed discussion by this Committee.

15.02 In paragraph 9.06, we listed the criteria which any technique of appraisal must satisfy before it can be adopted by the Department. They included tests of reliability, consistency and professional and public acceptability. We have proceeded on the assumption that the assessment methods applied to the items of cost appearing in COBA, although far from perfect, already satisfy these tests, and that the standard of performance which any other form of monetary valuation has to achieve is at least as but no more exacting than that achieved in the appraisal which underpins COBA.

15.03 We now apply those tests to the different techniques described in Chapter 14.

Actual Costs

15.04 The actual costs of mitigation or protection discussed at length in paragraphs 12.19 to 12.25 and 14.06 to 14.09 do not give rise to any difficulty except the practical problem of accurate identification and measurement. If they can be reliably estimated, they should form part of the costs, discounted as necessary, which are reflected in the scheme's Net Present Value. Similarly, where a public or private body or individual is relieved of expenditure by environmental improvements brought about by a road scheme, that constitutes a direct benefit, to be treated in the same way. The more expanded form of environmental assessment which we have recommended should elicit these costs and benefits and state them.

Shadow Costs

15.05 At the end of Chapter 14, we referred to another category of cost which, in our view, would rank as an obvious candidate for cost-benefit analysis: costs directly imposed upon polluters by Government, by way of a tax, on the "polluter pays" principle. Alternatively, we suggested that the Government might choose, as an act of policy, to measure the cost of environmental damage against a specially estimated "shadow price". That price would be the tax which polluters would have to pay if such a tax were levied. It might be measured by reference to the cost of clean-up, or in some other way. In any event, it too should be incorporated as part of the direct costs of every scheme which is considered as an option in the appraisal. It should also be related to Do-Nothing or Do-Minimum, and alternative policy options.

15.06 At present, the Government does not impose any tax of the type under discussion, and it has not yet formulated any shadow prices as a measure of environmental pollution. The arguments for doing so, endorsed in the White Paper, are powerful; but it is for the Government, and not us, to decide whether any such pricing mechanism can be introduced. We anticipate that some further research will be commissioned before this idea could be put into practice, although there is no barrier in principle to implementing it as an act of judgement immediately.

Revealed Values, Contingent Valuations and Stated Preferences

15.07 We also recommend that progress should be made with the development of revealed values (property values and travel costs) and those derived from contingent valuation and stated preference techniques, but in this instance on a step-by-step basis. There is a considerable body of research literature on these methods, and their limitations as well as their strengths are well known. They are summarised in Chapter 14.

15.08 In our view, the time has come to apply these techniques experimentally, to a sample of actual road schemes, corridors and strategic policy assessments. We do not recommend their immediate adoption into the formal appraisal itself because the Department has not yet had the practical experience which would give some guide as to whether, at any rate in the case of road schemes, the known criticisms could be overcome. Without that experience the Department will not be able to defend the results convincingly at a Public Inquiry. The first step should therefore be the setting-up of pilot projects in a cross-section of different types of schemes and policy studies, to include revealed values, contingent valuations and values derived from stated preference surveys. The presentation of the results of these pilot projects should include all the elements of the actual assessment (including the additional factors recommended in paragraphs 15.04 to 15.06) so that the balance of the assessment, taken as a whole, can be judged.

15.09 Revealed values should be tackled in a number of ways. Experienced property valuers, either District Valuers or private practitioners, should be asked in the first instance to estimate the impact on the values of residential properties of those effects to which we know property-owners are most sensitive: noise, vibration, visual intrusion, air pollution, severance and other localised, perceived effects. So far as possible, all such effects should be distinguished separately. Estimates will have to be given both of the improvement in values enjoyed by properties which gain in amenity and the depreciation in value of those which might suffer. Work on these lines should be carried out on all options, for schemes which are at different stages in their evolution, from Scheme Identification Study right up to the Public Inquiry Stage. Other changes in property value may also have to be examined too, such as those resulting from gains and losses in accessibility, where these are not already accounted for by time-savings.

15.10 In parallel with these professional valuations, surveys of the same schemes should be carried out, of the type now familiar in academic studies of revealed values, in order to see whether there is any correlation between these two approaches. This is an important test of the reliability of revealed values, for obvious reasons. A reasonably close relationship between two sets of data gathered in different ways will substantially increase confidence in the techniques.

15.11 It may be that the figures will be shown to be sufficiently robust to justify their inclusion in the calculations of Net Present Value. It may be found that one type of scheme, for example an urban scheme to be built in a densely-populated neighbourhood, lends itself more readily to this kind of treatment than others; that the technique is more useful for some types of effect than others; and that there are some types of effect (such as noise) which can be expressed in standard figures, whereas others will always have to remain specific to their particular scheme. We wish to suspend final judgement on these points until some clear results are obtained.

15.12   As to contingent valuations and stated preferences, some of the values in COBA - the value of non-working travelling time and the value of fatal accidents, for example - are already derived from these techniques. We would therefore encourage the further extension of both of these techniques into environmental assessment, again on an experimental basis. Here too, we suggest that local and perceived impacts should be taken first. The Department should venture into difficult areas only when the more simple cases return results which command confidence.

Traffic Forecasting and COBA

15.13   Although our present terms of reference do not extend to a review of traffic forecasting methods or of COBA we have had, throughout our study, constantly to refer to them because of the central position which they occupy in our system of appraisal. We have found that they are at some relevant points weak or incomplete, and that these features have a direct bearing upon the reliability of our environmental assessment, the way in which it should be carried out, and the form in which it should be presented. We therefore wish to comment on these particular points.

15.14   First, the traffic forecasts upon which both COBA and the environmental assessment are founded are based upon a fixed trip-matrix. That is to say, as we stated in paragraph 4.02, "high growth" and "low growth" forecasts are made of the likely future pattern of trips based upon those using the existing network. Those trips are then re-assigned to any new network which is proposed. No attempt is made to predict the new total volume and pattern of travel which might be induced by the new network and associated land-use changes. It is well known that if such induced travel were to occur, leading to a level and pattern of traffic different from that used in the appraisal of the scheme, the calculations of time, money and environmental impacts could be distorted and even be perverse.

15.15   SACTRA has been asked by the Government (see our terms of reference in Annex 2) to give advice on this matter. This will be our next task. Until we can predict these responses with more accuracy, there is the ever-present risk, which can vary substantially from scheme to scheme, that all effects, economic and environmental, will be inaccurately assessed.

15.16   Secondly, so far as COBA is concerned, we have already drawn attention, in paragraphs 14.11 to 14.15, to a potential hidden bias which tends to favour considerations of value for money against environmental quality in one particular case. Moreover, the "reductionist" character of COBA, where all the monetised effects are reduced to a single indicator (the Net Present Value) can conceal more than it reveals.

15.17   Thus, if the economic outputs of disparate environmental effects are introduced into COBA, it will be of the greatest importance to counter-balance, elsewhere in the assessment, the reduction of such effects to a single Net Present Value.

15.18 This potential requirement will be met if, in the Environmental Assessment Reports and the statutory Environmental Statements, discussed in Chapters 11 and 12, there is a clear and full presentation of the impacts in the disaggregated physical units in which they have been measured. As the ambit of trade-offs between effects widens, and more disparate effects are brought in, there will be a greater need to ensure that the "raw material" used in the process is explicitly stated and fully understood.

Postscript

15.19 Finally, it is clear that, for the foreseeable future and quite possibly in perpetuity, it will not be possible to value all the elements of environmental costs and benefits in monetary terms. However many environmental impacts are (or can be) monetised, others will remain in physical units and the remainder as qualitative descriptions. This is particularly true in the case of effects which are more removed from those which directly affect people, and are not perceived by them. Further, even in those cases where monetisation is achieved, money values will not tell the full story. Thus, qualitative description of the impacts will always be important in appraisal.

15.20 For this reason, the method of presentation and scope of information will remain important too. Our own views on this matter are now to be found in Chapters 11 and 12 of this Report, and paragraph 15.18. The decision on any scheme will always be an exercise in political judgement in the end, but the quality of that decision is critically dependent upon the quantity, quality, and accuracy of the material upon which it is based.

PART VI

CONCLUSIONS AND RECOMMENDATIONS

CHAPTER 16
CONCLUSIONS AND RECOMMENDATIONS

Introductory

16.01 Since the Department of Transport's procedures for the assessment of the economic and environmental effects of road schemes were first established, scientific and general understanding of the effects of roads and traffic on the environment have been greatly enhanced. Also, the distinction between "economic" and "environmental" effects has been called into question (paragraph 1.08).

16.02 The environmental effects of road-building and road-use are now seen to be wide-ranging and complex, and they are difficult to classify (paragraph 2.04). The allocation of impacts to different levels of decision-making - strategic, regional and local - is one method of classification which has much to commend it, but it requires careful judgement (paragraphs 2.05 and 2.06 and Figure 2).

16.03 In order to ensure that proper account is taken of all effects, the appropriate environmental assessments must underlie every stage in the hierarchy of decisions, from the making of national and regional policy downwards. Different effects cannot be neatly compartmentalised, or allocated exclusively to one particular point in that hierarchy. Each level interacts with the other, and the decision process has to be an iterative one (paragraph 2.08), employing a wide range of expertise drawn from various disciplines (paragraph 2.10).

16.04 Effects may also be divided into short-, medium- and long-term; those which are perceived and those which are not; and those which bear closely, more remotely, and not at all, upon human beings. Each requires its own method of assessment (paragraphs 2.11 - 2.14).

16.05 An effective system of measurement must be capable of capturing all effects, and give them coherence with policies and decisions made at all levels in the planning of our transportation system (paragraphs 2.15 and 9.02).

16.06 There is a well-established and orderly procedure for the development of road-schemes from early planning to final decision, and any method of assessment must fit into this procedure (paragraph 3.10).

Good Practice

16.07 Scheme assessments should in every respect be guided by explicitly stated policy objectives, which will provide a series of reference points against which competing options can be judged (paragraph 9.03).

16.08 Environmental assessment on a scheme by scheme basis alone will not take account of all effects. There is a need for a strategic level of assessment (paragraph 9.04).

16.09 New Roads may enable additional traffic to be attracted to the network. In such cases, unless this is estimated, assessment of the environmental impacts may be perverse (paragraph 9.04a). The effect on speeds is also important, as very low speeds, and high speeds, produce higher fuel consumption per mile. There may be non-fuel environmental benefits from speed reductions, (paragraph 9.04b). Transport interacts with land use, so that the environmental impacts of policies used to reduce the need for car journeys and encourages public transport, walking and cycling also need to be assessed (paragraph 9.04c).

16.10 All factors should be given their due weight in the formal procedures which influence how much road building there should be, where it should be located, and what design standards should be applied (paragraph 9.05).

16.11 An appraisal structure must be devised which will be adequate in geographical extent and timescale, and in its consideration of the combined and cumulative impacts of several schemes and polices (paragraph 9.05).

16.12 The system of environmental assessment which we need must therefore derive from explicitly stated environmental policy objectives. It must be comprehensive; based on real evidence; practical; capable of consistent application; and command acceptance by professionals and the public. The resources and time necessary to operate it should not be out of proportion to the importance of the issue involved, or greater than the cost of the outcome. The form of reporting must be clear and non-technical. The value judgements and technical assumptions made, and the methods used, should be explicit, open to scrutiny, challenge and possible revision (paragraph 9.06).

16.13 Present assessment techniques are better equipped to deal with scheme-specific, locally perceived effects than strategic or long-term effects. There has in the past been less examination of the environmental consequences of roads policy as a whole (paragraph 10.02).

16.14 There is a shift in Government thinking towards more strategic environmental considerations which SACTRA supports and wishes to see developed (paragraphs 10.03 - 10.05).

16.15 Implementation of this new approach starts at the highest level with the setting of objectives or targets in relation to major impacts, and establishing procedures for monitoring and reporting progress towards their achievement (paragraph 10.06).

16.16 The procedures for the initiation, design and approval of trunk roads should relate in both timing and content to those for land-use and transportation planning. If due consideration of these important matters is not given at the outset, they are likely to be lost altogether when the somewhat narrower choices between different schemes at a more local level have to be made. This implies some formal assessment of environmental impacts at the regional level of a kind which we propose in Chapter 11 (paragraph 10.08).

16.17 Regional and local objectives come into play subsequently; but the planning process is iterative, and there will be continuous overlap. The appraisal of individual road schemes will achieve a proper focus only if it is carried out within a framework of national, regional and local environmental policies and objectives working in combination (paragraph 10.09).

16.18 The early stages of planning and design are of great importance. Existing procedures applicable to Route and Scheme Identification Studies should be codified more thoroughly. No scheme should be admitted into the Roads Programme until its performance against strategic policies has been assessed and reported, at least in outline (paragraph 10.10).

16.19 Similarly, where regional and more local environmental objectives have been adopted by local planning authorities which are relevant to roads, routes and schemes should be tested against these objectives also, at the Identification Stage. The results of this early appraisal should be available at the Public Consultation and Public Inquiry stages, so that the public may be satisfied that these matters have been fully taken into account, and to allow the results to be subjected to public scrutiny (paragraph 10.10).

16.20 After a scheme has entered the Roads Programme room should prudently be left within the appraisal process to ensure that differences between schemes which significantly affect national objectives are properly picked up. Strategic effects may be important in choosing between options (paragraph 10.11).

The Timing of Assessment

16.21 The statutory Environmental Statement required by EC Directive 85/337 and the Highways Act 1980 should be preceded by two formal non-statutory Environmental Assessment Reports, prepared (i) before a scheme enters the Roads Programme and (ii) before options are presented for Public Consultation (paragraphs 11.01 - 11.03).

16.22 These Reports will successively deal with wider national and environmental issues, corridor and regional effects, and the local, detailed considerations arising out of scheme-options. The assessment reports should become increasingly detailed in coverage, such that the environmental assessment of a route-option should be readily developable into an Environmental Statement if that option finally goes forward to Public Inquiry (paragraph 11.02).

16.23 The "Environmental Assessment Report - Roads Programme Entry" must be available to those who subsequently undertake the principal work of environmental assessment and a summary of its main contents will be reported in the "Environmental Assessment Report - Route Options". This latter report will be a document which is available to the public for the purposes of public information and consultation and will be available for discussion at the Public Inquiry. With respect to it, and to the statutory Environmental Statement which is produced at Stage c (see paragraph 11.03), copies should be made available to those who are most directly affected by the proposals. Other interested parties should be able to purchase copies of these documents on the basis of its price being related to cost recovery (paragraph 11.05).

Content of the Assessment

16.24   The wide terms of the EC Directive invite an environmental appraisal in the fullest sense. Every significant effect, direct and indirect, must be reported. There is no bias in the discussion towards any particular point on the global - national - regional - local spectrum. Properly designed, an Environmental Statement of this type is the flexible document best equipped to meet the several criteria set out in paragraph 9.06 (paragraph 11.06).

16.25   The statutory Environmental Statements produced on publication of the finally Preferred Scheme need to be considerably expanded and refined, and more sensitively attuned to the special features of each proposal (paragraphs 11.08 - 11.12, 11.15 and 11.18).

16.26   The practice of including the Framework derived from the MEA within the text of an Environmental Statement should be abandoned (paragraphs 11.13 and 11.14).

16.27   Further, assessment should no longer be tied to the six interest Groups identified in the MEA (paragraph 11.15).

16.28   The suggestion made in the SACTRA 1986 Report that the Framework method of presentation should be replaced for all purposes by an Assessment Summary Report is endorsed (paragraph 11.19).

16.29   The non-technical summary at the end of the Environmental Statement may provide some of the text for the Assessment Summary Report (paragraph 11.20).

16.30   It may still be appropriate to present comparisons between different effects in a tabular form in the Assessment Summary Report, but any suggestion that there is a mandatory format to which all such Reports must adhere should be avoided (paragraph 11.21).

Guidance on the Assessment

16.31   The existing MEA should be replaced by a new Environmental Assessment Manual which will concentrate exclusively on environmental effects and contain all relevant advice. The document could be presented in a form where sections could be replaced from time to time as advice develops (paragraphs 12.03 and 12.04).

16.32   The scope and content of the guidance which the new Manual gives should be greatly enlarged in comparison with the MEA and its technical guidance should be kept under review, particularly regarding the use of specific expertise (paragraphs 12.04 - 12.07 and Figure 4).

16.33 Specific additions to the existing checklist should include the effects of land-take; loss of open space; water pollution; vibration; and a general heading for health and stress. The method of selection of significant effects is as important as the description of them (paragraphs 12.08 - 12.14).

16.34 The recommendations contained in the SACTRA 1986 Report on the inclusion of night-time noise, and the social and environmental effects of blight and development potential were accepted, but have not yet been incorporated into the MEA (paragraph 12.12). For inter-urban schemes View from the Road must remain an important factor. Driver Stress and Effects on Pedestrians and Cyclists do not seem to fit logically into the rest of MEA's list of eleven topics (paragraph 12.13).

16.35 However, assessment does not consist simply of running through a checklist (paragraph 12.10).

16.36 We recommend the redefinition of the grouping of the generic environmental impact headings as currently listed in Part B of the MEA, so that areas of particular concern, such as effects on natural resources, socio-economic effects or strategic effects, are clearly covered in the assessment process (paragraph 12.15).

16.37 The new Manual must contain guidance on the assessment of strategic and regional as well as local effects (paragraphs 12.16 - 12.18).

16.38 Environmental assessment should include a systematic analysis of the possibility of preventing, mitigating or off-setting adverse environmental effects. Where environmental assets are threatened, the cost of carrying out such measures, wherever they are feasible, should be estimated and reported, even if they are not expected to be carried out in fact (paragraphs 12.20 - 12.25). See also paragraphs 16.46 and 16.48.

16.39 Paragraph 2.6.1 of HD 18/88 should be widened to include the possibilities of preventing or offsetting a significant environmental effect as well as simply reducing it. We have in mind such possibilities as the relocation or the re-creation of environmental assets which may be significantly affected and are of a kind amenable to measures of this sort (paragraph 12.19).

Monetary Valuation of Environmental Effects

16.40 We concur with the earlier Reports of ACTRA and SACTRA that a formal method of appraisal is required to furnish decision-makers with all relevant information in a form which will enable them to give due weight to all the disparate effects of competing proposals. Such a method is a tool in decision-making and not a substitute for it (paragraph 13.04).

16.41 The use of a common unit to measure impacts of a different character will simplify the balancing exercise which has to be carried out (paragraph 13.05).

16.42 Cost-benefit analysis and the use of money-values has advantages over other comparative methods (paragraph 13.07); but there are limitations and weaknesses which must be acknowledged (paragraph 13.08).

16.43 There are some environmental effects which cannot sensibly be valued (paragraph 13.08); but there is no legitimate objection of principle to valuing those effects which can be (paragraph 13.11).

16.44 Any method of monetary evaluation which is used must satisfy the criteria specified in paragraph 9.06 of this Report (see paragraph 16.12), and any bias against non-monetised effects must be avoided (paragraph 13.11).

16.45 Effects which cannot sensibly be monetised include those which may cause social catastrophe or the loss of unique or sacrosanct assets; and cumulative or irreversible effects which fall on future generations (paragraphs 14.02 - 14.05).

16.46 Where actual costs are incurred by any agency or person, for example in mitigating, off-setting or cleaning-up the effects of damage or pollution, they should be included in the cost-benefit analysis where they can reasonably be calculated (paragraphs 14.06 - 14.09).

16.47 The cost of acquiring land which is currently included in the COBA program may be a serious underestimate of its social value if the land is subject to severe restrictions on development (paragraph 14.11 - 14.15).

16.48 The costs of mitigation, clean-up and protection will always be relevant to decision-making, even if they are not expected to be incurred in fact (paragraphs 14.16 and 14.17).

16.49 Values revealed by behaviour, for example increases or decreases in the price of houses following road construction, may be a valid measure of the effect of certain local and clearly perceived effects (paragraph 14.18).

16.50 Such effects can be evaluated by a combination of professional estimation by qualified property-valuers and statistical analysis (paragraphs 14.19 - 14.22); but these techniques have known limitations and weaknesses (paragraphs 14.24 - 14.26).

16.51 The "travel cost" method is another type of analysis based on actual behaviour but we doubt whether the results obtained by this technique, in its present state of development, could be converted into economic value with any degree of confidence (paragraph 14.27 - 14.29).

16.52 Individual values for certain environmental assets and detriments can also be ascertained expressly by survey methods. The best techniques (contingent valuation and stated preference) are well thought of, despite known limitations and weaknesses (paragraphs 14.30 - 14.39).

16.53 Using the results from revealed values and statements of preference raises problems of equity and distribution. There are difficult choices to be made between using directly observed market values and standard values. There are arguments for using standard values for impacts of general application such as noise reduction which are as strong as they are for time and accident savings. If, however, there is to be a development from the case by case approach to standardisation, a substantial research programme will be required (paragraph 14.40).

16.54 For strategic impacts which are not currently perceived by individuals and need to be dealt with at a social level, there are techniques of actual taxation or "shadow pricing" for translating policy targets and constraints into money values (paragraph 14.41 - 14.49).

16.55 The standard to be applied for the inclusion within cost-benefit analysis of values for environmental effects should not be more exacting than that applied to values already included (paragraph 15.02).

16.56 Actual costs and benefits resulting from environmental change should always be included in cost-benefit analysis whenever they can be reliably ascertained (paragraph 15.04).

16.57 Taxes introduced for reasons of environmental policy should also be included, as soon as the Government introduces them and should not be "netted out" as a transfer payment in cost-benefit analysis (paragraph 14.45, 15.05 and 15.06). Where these taxes are not introduced, calculation of their notional level can still be used in assessing the shadow costs of macro-environmental impacts (paragraphs 15.05 and 15.06).

16.58 Pilot studies should be carried out, in relation to a representative cross-section of schemes and policies, to judge the reliability of revealed values, contingent valuations and stated preference techniques in the specific area of transport-planning (paragraphs 15.07 - 15.12).

16.59 The results of these studies should be amalgamated into the actual assessment being carried out into the schemes in question, (including the matters referred to in paragraph 16.57) to test their effect on the balance of assessment overall (paragraph 15.08).

Traffic Forecasting and COBA

16.60 Some of the techniques used in current traffic forecasting methods and COBA may distort the results of any environmental assessment and merit early consideration (paragraph 15.13 - 15.17).

Presentation of the Assessment

16.61 The complex and diverse nature of the environmental and economic effects of road-building will always require the full description and presentation of each effect in its disaggregated form, separately stated and measured, notwithstanding any improvement in techniques for monetary valuation (paragraphs 15.18 and 15.20).

ANNEX 1

EVIDENCE RECEIVED

We would like to thank the following who have submitted evidence to us in the course of this work:

Government Departments

- Department of the Environment
- Department of Transport
- Her Majesty's Treasury

Local Authorities

- Avon County Council
- Bradford Metropolitan Borough Council, Directorate of Enterprise and Environment
- Cheshire County Council, Highways Services Department
- Cleveland County Council, County Surveyor and Engineer Department
- Coventry, City Engineers Department
- Dorset County Council, Transportation and Engineering Department
- Gloucestershire County Council, County Surveyor's Department
- Hertfordshire County Council, Department of Highways
- Lancashire County Council, County Surveyor and Bridgemaster's Department
- Lincolnshire County Council, Highways and Planning Department
- London Borough of Hammersmith and Fulham, Transport Networks Group, Development Planning
- London Borough of Merton, Development Department
- London Borough of Richmond Upon Thames, Technical Services Department
- North Yorkshire County Council, Highways and Transportation Department
- Oxfordshire County Council, Surveyor and Engineer's Department
- Rochdale Metropolitan Borough Council, Technical Services Department
- Somerset County Council, Surveyors Department
- Stockport Metropolitan Borough, Development and Town Planning Department
- Surrey County Council, County Engineer's Department
- West Sussex County Council, County Engineer and Surveyor's Department
- West Yorkshire Highways, Engineering and Technical Services Joint Committee (HETS)
- Wiltshire County Council, Department of Planning and Highways

Interested Organisations

- Association of Consulting Engineers
- Association of County Councils
- Association of District Councils
- Association of Metropolitan Authorities
- Automobile Association
- Bartlett School of Architecture and Planning, University College London
- Bioscan (UK)Ltd
- Bristol Polytechnic
- British Road Federation Ltd
- Cadw, Welsh Historic Monuments
- Civic Trust
- Cobham Resource Consultants

- Confederation of British Industry
- Council for the Protection of Rural England
- Countryside Commission
- Cyclists Touring Club
- Cyclists Touring Club, N Yorks District
- Economic Studies Group
- English Heritage
- Environmental and Transport Planning
- Environmental Assessment Services Ltd
- Freight Transport Association
- Halcrow Fox Associates
- Harrogate Conservation Volunteers
- Incorporated Society of Valuers and Auctioneers
- Institute of Road Transport Engineers
- Institute of Terrestrial Ecology, Natural Environment Research Council
- Institution of Civil Engineers
- Institution of Highways and Transportation
- Kirkpatrick and Partners
- L G Mouchel and Partners
- Landscape Advisory Committee
- Landscape Institute
- London Cycling Campaign
- London Regional Transport
- Metropolitan Transport Research Unit
- Mott MacDonald
- Nathaniel Lichfield and Partners
- National Trust
- National Association for Environmental Education
- Nature Conservancy Council
- Noise Abatement Society
- Open Spaces Society
- Open University, Faculty of Technology
- Oxford Economic Research Associates Ltd
- Polytechnic South West, Institute of Marine Studies
- Ramblers Association
- Road Operators Safety Council
- Royal Automobile Club
- Royal Commission on Environmental Pollution
- Royal Institution of Chartered Surveyors
- Royal Town Planning Institute
- Royal Society for Nature Conservation
- RPS Clouston
- Scott Wilson Kirkpatrick
- Scottish Development Agency
- Sir Bruce White, Wolfe Barry and Partners
- Sports Council
- Town and Country Planning Association
- Transport 2000
- University of Birmingham, School of Civil Engineering
- University of Glasgow, Department of Economic History
- University of Leeds, Institute for Transport Studies

- University of Salford, Civil Engineering Department
- Urban Economic Development Group
- W J Cairns and Partners

Individuals

- County Planning Officer and County Surveyor, Bedfordshire County Council
- Dr J Adams, Department of Geography, University College London
- Mr D Bicknell, Director Of Engineering and Transportation, Royal Borough of Kingston upon Thames
- Mr G D Clarke, Walton-on-Thames, Surrey
- Mr M Draper, Chief Engineer, London Borough of Brent
- Mr M Elias
- Mr A Goldstein, Edenbridge, Kent
- Dr P Hawkins, Edinburgh
- Professor C I Howarth, Department of Psychology, University of Nottingham
- Mr H McClintock, Nottingham
- Prof D Pearce, Department Of Economics, University College London
- Mr D W Rennie, Department of Technical Services, Humberside County Council
- Mr J H Scott-Park, Alexandria, Dunbartonshire
- Mr M J Smith, Department of Mathematics, University of York
- Dr S P Wolff, Department of Clinical Pharmacology, University College London and Middlesex Hospital Medical School

Other evidence received

We would also like to express our gratitude to those authors whose evidence was submitted to us in confidence and whose names are therefore not recorded here.

ANNEX 2

TERMS OF REFERENCE AND MEMBERSHIP

Terms of Reference

- to advise the Department on the evidence on the circumstances, nature and magnitude of traffic re-distribution, mode choice and generation, especially on inter-urban roads and trunk roads close to conurbations, and to recommend whether and how the Department's methods should be amended, and what if any research or studies could be undertaken

- to review the Department's methods for assessing environmental costs and benefits, in particular whether a greater degree of valuation is desirable, the appropriate scope and application of valuation, and suitable methods for deriving monetary values

- to continue to advise on any significant changes proposed in appraisal methods.

The present report is concerned with the second of the above three Terms of Referencce.

Membership of the Committee

Mr D Wood QC - Chairman
    Principal of St Hugh's College, Oxford

Mr R H Stewart - Vice Chairman
    Chairman of the Planning Division and Director of Travers Morgan Environmental Ltd within the Travers Morgan Group

Mr H J Wootton
    Chairman of Wootton Jeffreys Consultants Ltd

Dr P B Goodwin
    Director of the Transport Studies Unit, University of Oxford

Professor P J Hills
    Professor of Transport Engineering, University of Newcastle upon Tyne

Mr P J Mackie
    Senior Lecturer in Economics, University of Leeds

Miss A M Lees
    Chairman of ACRE, the Association of Rural Community Councils. Formerly Controller of Transportation and Development for the Greater London Council

Acknowledgement

We would like to acknowledge the assistance received from the Department's staff and in particular the following who were members of the Secretariat or who provided support work during the course of this work:

- Mr S Grayson
- Mr R Lamsdale
- Mr S C Laurence
- Mr J L Myers

ANNEX 3

THE MAIN STAGES IN THE PLANNING AND CONSTRUCTION OF A TRUNK ROAD

The following is a chronological list of main events, including a brief description with particular reference to environmental assessment. The numbers in brackets refer to the box numbers in the attached chart, which gives a more detailed breakdown of the stages in a scheme's development.

THE NEED (1):

Pressure from public, MPs, Local Authorities etc, monitoring by Regional Offices; prediction of problems in future; or national policy formulation (White Paper objectives). Need may be environmental concern (eg to reduce local noise and pollution), safety, or to improve journey times and reduce congestion.

ROUTE IDENTIFICATION STUDY (2):

These are occasionally carried out to examine if there is a strategic or corridor need which might have a feasible solution. Public transport solutions may be examined, intermodal surveys may be included. Examine current and future demands. Suggested alignments will be very approximate. Environmental assessment is likely to concentrate on current problems and identification of sensitive areas rather than detailed estimates of effects of new roads. Studies vary significantly in scope, content, and style, but there is a tendency to greater scope and detail.

SCHEME IDENTIFICATION STUDY (2):

May be a study of a particular section within a previous Route Identification Study, but more commonly a study of an independent, less extensive, scheme though still needing to take account of any other schemes or developments in area. Carried out by consultants or agent authorities on behalf of Regional Offices to examine whether an identified need is worth addressing, and whether solutions might be economically and environmentally acceptable. Alignments may be approximate. Detailed and costly surveys and forecasts unlikely, though DTp are trying to get more realistic estimates at this early stage, but a wide range of uncertainty is always likely. Assessment should include environmental effects of potential options (probably broad brush), existence of SSSIs etc, policy implications (eg on roads in National Parks), proximity to communities. No requirement for a formal MEA Framework, but Frameworks are sometimes produced. Usually an input from landscape architects (not Landscape Advisory Committee) but detailed advice limited due to lack of precise alignments. Where alignment options are more constrained greater precision may be possible, eg number of demolitions. Development effects usually only discussed in terms of local and national policies, without knock on effect in terms of traffic and environmental impact.

PROGRAMME ENTRY (3):

When Region's Operating Unit is satisfied that there is a prima facie case, scheme is put to HQ for consideration by Ministers. HQ specifies what application should cover, eg problem identification and objectives and how proposals address them, value for money, statement on significant environmental impacts. No requirement for formal Framework or COBA. Includes views of local authorities and results of any early consultation. Announcement of Ministers' approval for programme entry normally made in White Paper or Roads Report every two years. Criteria for these reviews reflect policy aims. Programme entry is announcement of Government's intention to progress towards building, provided further studies show scheme is economically and environmentally justified and that it eventually can be afforded; not a commitment to build.

CONSULTING ENGINEERS APPOINTED (4):

Detailed design and appraisal begins after Programme entry, with the issue of a scheme brief. This sets terminal points and objectives, notes specific problems, environmental constraints, types of solution to study, and types of traffic, economic and environmental assessment to be carried out. The first major decision stage is choice of options to put to Public Consultation; the relevant assessments are noted below.

INITIAL SURVEYS AND ASSESSMENTS PRIOR TO PUBLIC CONSULTATION (5-7):

The aim is for the Secretary of State to be satisfied that he would be prepared to build all options to be put to consultation, with the decision normally delegated to the Regional Director. (The Regional Director is a joint DOE/DTp official.) Necessary to carry out traffic surveys, produce and validate traffic model, produce traffic forecasts, economic and environmental assessments. Detail will be finer than prior to programme entry, but coarser than will subsequently be done before publication of orders. Formal appraisal stages are Local Model Validation Report and Forecasting Report (the Traffic Appraisal Manual sets standards and methods for both) and COBA. The Landscape Advisory Committee normally make a site visit, reporting on quality of existing landscape, identification of conservation areas etc, their views on landscape effects of options, any suggested improvements, and their own recommended option. There will be consultations with local authorities and statutory environmental organisations, eg English Heritage, Countryside Commission. The scheme's first Framework is then drawn up incorporating relevant information from all these sources and the environmental studies necessary to meet MEA's requirements. (See Framework example in Annex 4.) If any options are Net Present Value negative colour coding is required to determine where the authority lies for approval of putting the option to consultation.

TECHNICAL APPRAISAL REPORT (8):

This internal document draws together all the relevant appraisals to justify the options to be put to Public Consultation. Refers to extent to which problems are solved, objectives met, and the operational assessment, economic performance and environmental assessment. Description and comparison may use material of greater detail than appears in Framework.

PUBLIC CONSULTATION (9):

Purpose is to inform public that a road scheme is being considered to deal with a particular problem or objective, to indicate possible options and consequences, and discover public's views on the options and relative importance of expected consequences. Public Exhibition(s) present descriptions and assessments, generally in greater detail than Framework. (Framework is also provided, and separately provided to various interested groups.) Consultation pamphlets giving the essential details are distributed in the locality, and include a questionnaire for public to send in views. Local discussion groups may be organised.

FINALISE PRELIMINARY REPORT (10):

Revisions to Local Model Validation Report, COBA, and environmental assessment may continue to this point. Preliminary Report has three parts; the Technical Appraisal Report, the Report on the Public Consultation, and the Assessment Report which is a distillation of the other two. The Assessment Report makes a clear recommendation of a preferred route, with reasons.

ANNOUNCEMENT OF PREFERRED ROUTE (11):

By the Secretary of State on basis of Preliminary Report, assuming of course he accepts the recommendation. Protects the line and authorises start of detailed design work.

DETAILED SURVEYS, DESIGN AND ASSESSMENT (12);

Will reflect need for more precision as required in draft line orders. May need extensive additional surveys, environmental assessments etc if the selected option is revised as a result of consultation. Design will include environmental protection measures. Landscape Advisory Committee invited to give views on any route changes. Further consultations with public bodies.

ORDER PUBLICATION REPORT (15):

Sets out final scheme details, and is clearance document on all appraisal and design procedures, ie to ensure all the Department's requirements have been properly met, and approval for continued preparation. Includes a Framework incorporating revised assessments, and now to be of the type shown in Annex 5. Colour coding if Net Present Value negative.

PUBLICATION OF DRAFT LINE ORDERS (17):

Negotiations with those affected, with any revisions to COBA, Framework etc. The Environmental Statement, as required by the implementation of the EC Directive 85/337, is published at the same time where necessary.

PUBLIC INQUIRY (19):

A Statutory proceeding. Held unless, following negotiations, there are no significant objections to Draft Line Orders. (Detailed revision of economic and environmental appraisal may continue right up to Public Inquiry.) The Inquiry Inspector will have available to him the final Framework, all the relevant COBA and Traffic Appraisal manuals etc, detailed scheme assessment reports and design drawings, written submissions and proofs of evidence from objectors, counter objectors and the Department. Additional papers may be called for during the Inquiry and the Department may be called upon to do additional appraisals.

PUBLISH DRAFT COMPULSORY PURCHASE ORDERS AND SIDE ROAD ORDERS (21):

A further Inquiry may be necessary.

INSPECTOR'S REPORT(S) AND SECRETARIES OF STATE'S DECISIONS (23):

Inspector's Report will include findings on fact, conclusions, and recommendations. If he agrees the need, may still make recommendations for changed alignments, standards, footbridges, environmental protection etc. The Secretaries of State for Environment and Transport have joint responsibility, and will reach decisions in the light of the Inspector's Report and all other factors. Orders may be amended accordingly. A new inquiry may be necessary.

ECONOMIC REVIEW AND WORKS COMMITMENT APPROVAL (24):

A full economic review, incorporating any amendments, cost estimate revisions etc. will be made before approval for works commitment, invitation to tender, letting of contracts, and the start of construction.

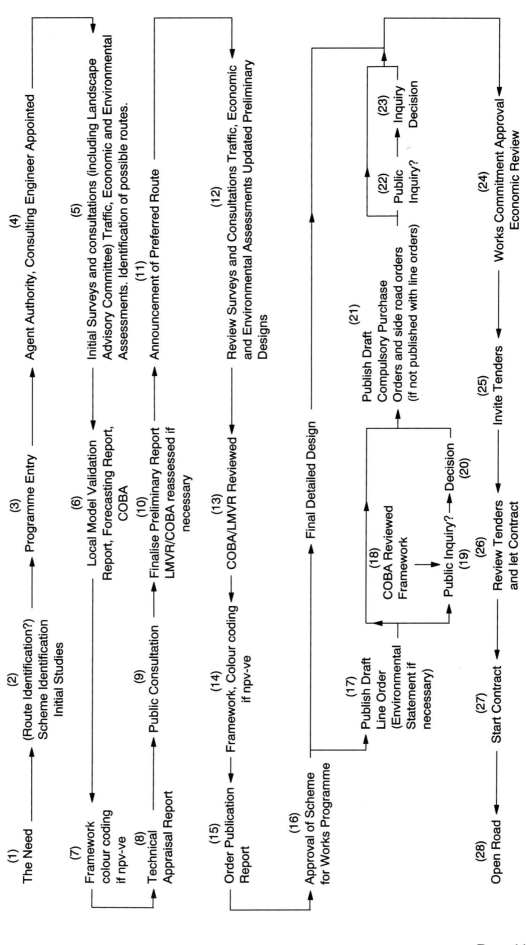

STAGES IN THE EVOLUTION OF TRUNK ROAD SCHEMES

ANNEX 4

MANUAL OF ENVIRONMENTAL APPRAISAL:

PART A SECTION 2 APPENDIX 1
PUBLIC CONSULTATION FRAMEWORK

Part A Section 2 Appendix 1

PUBLIC CONSULTATION FRAMEWORK Date Prepared: June 1983

Group 1: Travellers

Sub-Group	Effect	Units	Brown[1] High	Brown[1] Low	Blue High	Blue Low	Green High	Green Low	Red High	Red Low	Do Minimum	Comments
All vehicle travellers	Time savings	£m (PVB)	6.35	3.61	6.49	3.69	5.85	3.33	2.82	1.60	0	Note A, B and C each apply to the first three lines. A. Each column shows the improvement of the particular route over the 'Do-minimum' option. Hence the 'Do Minimum' entries are zeros. B. Present Value of Benefits (PVB) for a 30 year period from the expected date of opening and discounted to 1979 at 7% p.a. Negative figures are increased costs. These estimates are very preliminary at this stage and should be interpreted cautiously. C. It has been assumed that national average figures for vehicle occupancy and for accident rates and costs, will apply. The Red option assumes Alcester Road link will be constructed by the County Council
	Vehicle operating cost savings	£m (PVB)	−0.12	−0.10	−0.06	−0.05	+0.13	+0.11	+0.07	+0.06	0	
	Value of accident savings	£m (PVB)	0.29	0.20	0.25	0.17	0.25	0.17	0.20	0.14	0	
	Reduction in casualties:—											The figures indicate the probable total reduction in casualties over the whole of the 30 year assessment period if the national average rates and the distribution between groups apply to each alternative. They take no account of the safety implications of the detailed design of the new routes
	Fatal	Number	5	4	5	4	5	4	4	3	0	
	Serious	Number	48	33	42	29	42	29	33	23	0	
	Slight	Number	88	61	77	53	77	53	61	42	0	
	Traffic delays during construction		Slight		Slight		Moderate		Slight		Slight	
	Change in amenity		Reduction of traffic in town centre will improve safety		As Brown		As Brown		Partial reduction of town centre traffic will only slightly improve safety. Alcester Road will continue to be a hazard			
Pedestrians												Pedestrian/traffic conflict will increase with traffic growth. The construction of the M99, programmed for 1990, is expected to increase traffic and HGV's through the town

[1] Each scheme option should have a separate column.

Part A Section 2 Appendix 1

Group 2: Occupiers

Sub-Group	Effect	Units	Brown	Blue	Green	Red	Do Minimum	Comments
Residential	Properties demolished	Number	less than 5	less than 5	less than 5	None	None	The cost of property acquisition and development is included in Group 6
	Noise effects adjacent to new road	Number of houses within given distance of centre line						'Do Minimum' entry shows numbers of houses fronting the existing route from Low Rd Methodist Church to West of Horton
		0–50m	8	7	7	3		
		50–100m	20	17	17	0	(487)	
		100–200m	18	30	12	12		
		200–300m	32	14	28	10		
	Noise effects adjacent to existing roads	Number of premises experiencing at least a halving of the present traffic flow	400	380	380	290	0	A halving of traffic flow indicates approximately a 3dB(A)L_{10} (18 hour) noise reduction
	Visual obstruction		Severe for 9 houses	Severe for 4 houses. Significant for 3 houses	Severe for 5 houses	Slight for 8 houses	No change	
	Visual intrusion		Slight intrusion for houses in Horton Village	Slight intrusion for town suburb	Moderate intrusion where River crossing is visible from town	None	No change	
	Severance		Slight increase in community severance in Old Town. Slight reduction in West District	Moderate reduction of severance in West District. Severe new severance in the South	Reduced in West District. As Blue	Reduced in West District As Blue	Deteriorating situation will cause difficulties for pedestrians and cyclists leaving adjacent housing estates	
	Disruption during construction		Likely disruption to small number of houses	As Brown	As Brown	Likely disruption to a group of houses near road	None	
Industrial Premises	Noise effects adjacent to new road	Number of premises within 300m of centre-line 1	3	3	3	1	(4)	'Do Minimum' entry shows number of premises fronting existing route
	Noise effects adjacent to existing roads	Number of premises experiencing a halving of traffic flow	2	2	2	2	0	
	Visual obstruction		Slight	Slight	Significant	No effect	No change	
	Severance		None	None	None	None	No change	
	Disruption during construction		Slight	Slight	None	None	Slight	
Commercial premises								
a) Office Accommodation	Noise effects adjacent to new road	Number within 300m of centre-line 1	0	0	0	1	(12)	'Do Minimum' entry shows offices fronting existing route
	Noise effects adjacent to existing roads	Number experiencing halving of traffic flow	8 Daily occupancy approx. 600 people	8 Daily occupancy approx. 600 people	8 Daily occupancy approx. 600 people	7 Daily occupancy approx. 100 people	0	
	Visual obstruction		None	None	None	Slight	No change	
	Severance		None	None	None	None	No change	
	Disruption during construction		None	None	None	Slight	None	
b) Shops	Noise effects adjacent to new road	Number within 300m of centre-line	2	2	2	2	(35)	'Do Minimum' entry shows number of shops on existing road
	Noise effects adjacent to existing roads	Number experiencing halving of traffic flow	29	29	29	26	0	
	Visual obstruction		Slight	Slight	Slight	Moderate	No change	
	Severance		Slight	Slight	Moderate	Moderate	No change	
	Disruption during construction		Minor disruption to shops access	As Brown	As Brown	As Brown	Increasing difficulties of access for shoppers as traffic increases

NOTE:
1 The example is for a rural area. In built-up areas where substantial screening is likely it will normally be sufficient to list only those properties fronting or backing on to the route. In such a case a note to this effect should be included in the comment column.

Part A Section 2 Appendix 1

Group 2: Occupiers (Continued)

Sub-Group	Effect	Units	Brown	Blue	Green	Red	Do-Minimum	Comments
Schools and Hospitals								
a) Barchester Primary School (220 pupils in 1971)	Reduction of noise		Relief to 1 class room	As Brown	As Brown	Insignificant change	Noise increase as traffic flow increases.	
	Visual obstruction		Moderate obstruction to 4 class rooms	As Brown	As Brown	As Brown	No change	
	Severance		Slight Improvement	Slight Improvement	Slight Improvement	No change	No change	
	Disruption during construction		Moderate for 6 months	None	None	None	None	
b) Horton Cottage Hospital (40 beds, Accident Unit and Outpatients Dept. open week days only)	Noise increase	Within 300m of centre line of new road	No	No	Yes, Outpatients Dept.	Yes, Outpatients Dept.	No	
	Visual obstruction		None	None	Slight	Slight	No change	
	Severance (New)		None	None	None	None	None	
	Disruption during construction		None	None	Slight	Slight	None	
Farming	Land take	Number of Farms affected	12	15	11	19	0	
		Hectares of land Grade 2	5	4	12	23	0	Based on MAFF Classification. Compensation included in Group 6
		Grade 3A	13	22	21	18	0	
Open Space								
a) Horton Golf Course (area 46 hectares)	Land take	Hectares	0	0	6.0	0	0	Effects of Users appear in Group 3
b) Martin Public Park (area 30 hectares)	Land take	Hectares	2.8	0	0	0	0	Effects of Users appear in Group 3. Exchange Land Certificate required
Public Buildings								
a) Low Road Methodist Chapel (area 2·8 hectares)	Land take	Hectares	0	1·3	0	0	0	Effects on Users appear in Group 3. Home Office Certificate required to exhume remains

Group 3: Users of Facilities

Sub-Group	Effect	Brown	Blue	Green	Red	Do-Minimum	Comments
a) Town Centre Shoppers High St/ Market Street (100,000–150,000 Shoppers per week in 1980)	Reduction of vehicle/pedestrian conflict	Reduces and diverts traffic sufficient to allow pedestrianisation	As Brown	As Brown	Slight reduction of traffic giving little benefit	No traffic relief: traffic increase with time	Numbers of shoppers based on County Council Shopping Study (1979) and Chamber of Commerce (1980)
b) Users of Community Centre including Public Library (Number not yet known)	Changes in traffic	Moderate reduction in traffic	Moderate reduction in traffic	As Blue	Slight reduction in traffic	No traffic relief: traffic increase with time	
c) Users of Horton Golf Course (360 members in 1971)	Reduction in amenity through landtake	No effect	Reduced to 16 holes	No effect	No effect	No effect	
d) Users of Sailing Club (98 members 1971)	Reduction in amenity (visual intrusion, sailing conditions etc.)	8m embankment effectively prevents sailing on upper reach	As Brown	As Brown	No effect	No effect	
e) Horton Hunt (240 members in 1970)	Severance	Severe effect on traditional fox runs north of town because they become unuseable	As Brown	As Brown	Moderate severance of fox runs to South and West of town	No effect	
f) Martin Public Park (Numbers of users not yet known)	Reduction in amenity due to landtake	10% reduction in size including one football pitch	No effect	No effect	No effect	No effect	Largest of town's three public parks and playing fields
g) Low Road Methodist Chapel	Severance, Noise, Amenity	No effect	Severed from main part of town, frontage access affected	As Blue	No effect	No effect	Only Methodist Chapel in town

Page 121

Part A Section 2 Appendix 1

Group 4: Policies for conserving and enhancing the area

Policy	Authority	Interest	Brown	Blue	Green	Red	Do-Minimum	Comments
a) To Protect the Hill Street Outstanding Conservation Area	Dartshire CC Barchester DC	Improvement of the environmental quality of the conservation area, and reduction of pedestrian/vehicle conflict	Moderate reduction in traffic	Sufficient traffic is diverted to allow pedestrianisation	As Blue	No reduction	Traffic levels will increase with time	DOE designated area as Outstanding in 1976
b) To protect Listed Buildings	DOE Dartshire CC Barchester DC	Effect on Wattle Hall, a Grade II listed building	No effect	New road is 300m from Hall	No effect	No effect	No effect	Listing is based on interior fittings and ceilings
c) To restore derelict land in the Avon Valley	DOE Dartshire CC	Restoration of abandoned gravel pits	Part required for road scheme, remainder could be used as spoil tip	As Brown	As Brown	If used as spoil tip the material would be transported through town centre	No effect	See also Dartshire CC policy on Country park
d) To create a Country Park Leisure Centre adjacent to River Avon west of Barchester	Proposed and supported by:– Dartshire CC Barchester DC Countryside Commission Opposed by Horton PC	Ability to create park adjacent to river bank and including abandoned gravel pits	Reduces the area and scope to create park	Would be severed by road on embankment	As Blue	Would not affect the creation of the park but would increase heavy goods traffic from north going along park roads	Allows scheme to proceed	The park appears as a policy in both the Structure and Local Plans and has the potential for grant aid from the Countryside Commission
e) To relocate Cattle Market and Slaughter house	Barchester DC	Preservation of new site	No effect	No effect	No effect	Takes land allocated under local plan	No effect	Local Plan policy

Group 5: Transport, development and economic policies

Policy	Authority	Interest	Brown	Blue	Green	Red	Do-Minimum	Comments
Transport								
a) To improve trunk roads to ports	Department of Transport	Ease of access from manufacturing centres to the port	Big improvement	Big improvement	Big improvement	Some Improvement	Increasing delays expected	White Paper on Road Policy 1980
b) To relieve local traffic problems in Barchester	Dartshire CC	Removal of through traffic	Most through traffic removed	As Brown	As Brown	Little benefit	Increasing traffic delays	Dartshire CC are highway authority
c) To Improve safety and to upgrade the London Camelot railway line	British Rail	Removal of Heton level crossing	Removes crossing	Removes crossing	Crossing remains	Crossing remains	Crossing remains	
d) To maintain River Avon navigation	British Waterways Board	Temporary effect of bridge construction on navigation	A new bridge to be constructed	As Brown	2 new bridges to be constructed	No effect	No effect	
e) To maintain a viable rural bus transport system in South Dartshire	Dartshire CC Bus Operators	Removal of congestion	Removes congestion	As Brown	As Brown	No effect	Existing traffic problems will increase	
Development & Economic								
a) To develop Barchester as a regional shopping centre	Dartshire CC Barchester DC	Improved accessibility to Barchester by removal of through traffic	Removes through traffic	As Brown	As Brown	No benefit	Current traffic problems worse	County Structure Plan and Local Plan policy
		Improved amenity in shopping area of High Street	Pedestrian/vehicle conflict reduced	Permits pedestrianisation	Permits pedestrianisation	No benefit	Current problems increase	

Part A Section 2 Appendix 1

Group 5: Transport, development and economic policies (continued)

Policies	Authority	Interest	Brown	Blue	Green	Red	Do-Minimum	Comments
b) To limit growth in South Dartshire and encourage new employment and housing in villages of Scapton Haydon and Wettering	Dartshire CC	Effect on the rural northern sector of the County	Improves access to Scapton, Haydon and Wettering	As Brown	As Brown	No effect	No effect	County Structure Plan policy
c) To safeguard indentified commercially workable gravel resources in the Avon Valley	DOE Dartshire CC Barchester DC	Gravel beds west of Barchester	No effect	No effect	3 hectares affected	No effect	No effect	County Structure Plan policy
d) To permit further growth in Horton Village (increase approx. 400 houses)	Proposed by Dartshire CC Opposed by Barchester DC and Horton PC	Preservation of expansion area	Will encourage growth as dormitory village	No effect	No effect	No effect	No effect	It is likely that Horton Village will eventually develop up to the line of the new road
e) To encourage all existing non-conforming industry to relocate and all new industry to locate on the Barchester Industrial Estate	Proposed by Barchester DC Opposed by Dartshire CC	Effect on access to Industrial Estate	No effect	Improves access	No effect	No effect	No effect	Both the District and County Councils favour concentration of new and non conforming industry on industrial estates, but Dartshire would prefer growth at Blaydon City rather than Barchester. Non conforming industry is that not compatible with the general land use in the area

Group 6: Financial Effects

Sub-group	Interest	Units	Brown		Blue		Green		Red		Do-Minimum	Comments
			High	Low	High	Low	High	Low	High	Low		
Department of Transport	Construction costs **1**	£m (PVC)	6·52	3·71	6·68	3·81	6·23	3·61	3·09	1·80	0·4	Costs are discounted from year of expected expenditure to 1979 at 1979 prices. Cost quoted in other documents may well be on a different price basis and discounted to a different base year (PVC = present value of cost, PVB present value of benefit, NPV = net present value)
	Land costs	£m (PVC)									0·1	
	Total cost	£m (PVC)									0·5	
Dartshire CC	Construction costs **1**	£m (PVC)									0	
	Land costs	£m (PVC)									0	
	Total cost	£m (PVC)									0	
			6·2		4·9		4·1		3·9			
			1·0		1·0		1·0		0·8			
			7·2		5·9		5·1		4·7			
			0		0		0		1·0			
			0		0		0		0·2			
			0		0		0		1·2			
Total quantified monetary benefit		£m (PVB)	–0·18		+1·28		+1·63				0	Includes savings in time, vehicle operating costs and accidents. Taken from Group 1
			6·52	3·71	6·68	3·81	6·23	3·61	3·09	1·80		
Net present value compared to do-minimum		£m (NPV)	–0·18	–2·99	+1·28	–1·59	+1·63	–0·99	–2·31	–3·60	0	

Note:
1 Construction costs should include preparation and supervision costs.

ANNEX 5

MANUAL OF ENVIRONMENTAL APPRAISAL:

PART A SECTION 2 APPENDIX 2
PUBLIC INQUIRY FRAMEWORK

Part A Section 2 Appendix 2

PUBLIC INQUIRY FRAMEWORK

Date Prepared: June 1983

Group 1: Travellers

Sub-Group	Effect	Units	Modified Blue		Modified Green		District Council		Do Minimum [1]	Comments
			High	Low	High	Low	High	Low		
Car Users	Time savings	£m (PVB)	4·63	2·36	3·80	2·16	4·19	2·38	0	Notes A, B and C apply to the first nine lines
	Vehicle operating cost savings	£m (PVB)	−0·14	−0·12	+0·23	+0·20	0·00	0·00	0	A. Each column shows the improvements of the particular route over the "do-minimum" option. Hence the "Do Minimum" entries are zero.
Users of Light Goods Vehicles	Time savings	£m (PVB)	1·06	0·60	0·96	0·54	1·09	0·62	0	
	Vehicle operating cost savings	£m (PVB)	−0·05	−0·04	+0·03	+0·03	−0·06	−0·05	0	B. Present value of benefits (PVB) are for 30 year periods from the expected date of opening and discounted to 1979 prices at 7% pa.
Users of other Goods Vehicles	Time savings	£m (PVB)	0·96	0·54	0·42	0·24	0·13	0·08	0	
	Vehicle operating cost savings	£m (PVB)	−0·05	−0·04	+0·02	+0·02	−0·02	−0·02	0	
Bus Operators and passengers	Time savings	£m (PVB)	0·64	0·36	0·56	0·32	0·74	0·42	0	C. It is assumed that national average figures for vehicle occupancy and for accident rates and costs will apply.
	Vehicle operating cost savings	£m (PVB)	−0·01	−0·01	0·00	0·00	−0·02	−0·02	0	
All vehicle travellers	Value of accident savings	£m (PVB)	0·25	0·17	0·18	0·13	0·18	0·13	0	The figures indicate the probable total reduction in casualties over the whole of the 30 years assessment period if the national average rates and distribution between groups apply to each alternative. They take no account of the safety implications of the detailed design of the new routes.
	Reduction in casualties:—									
	Fatal	number	5	4	4	3	4	3	0	
	Serious	number	42	29	31	22	31	22	0	
	Slight	number	77	53	56	39	56	39	0	
	Driver stress		Low		Moderate		Moderate		High	
	View from road		Scenic		Agricultural		Agricultural		Residential	
	Traffic delays during construction	£m (PVB)	−0·02	−0·02	−0·02	−0·02	−0·28	−0·02	0	Figures are calculated using the same assumption on traffic composition as for travel benefits. No detailed survey has been undertaken
Pedestrians [2] (5000–8000 pedestrian movements per day will be affected)	Change in amenity		Pedestrianisation of Town Centre and removal of heavy traffic will improve the quality of the adjacent streets		As modified Blue		As modified Blue		A 20% increase in traffic will reduce amenity in the town centre	
	Safety		Segregation of pedestrians and vehicles will improve safety		As modified Blue		As modified Blue		With a 20% increase in traffic the danger of accidents to pedestrians in the town centre will increase	
	Severance (New)		Severe: Several footpaths diverted		Moderate: Pedestrians will have to use subway to reach hospital		As modified Green			

[1] This heading will be 'Do Nothing' for many schemes.
[2] If the effect on cyclists is thought to be important a separate sub-group should be included.

Part A Section 2 Appendix 2

Group 2: Occupiers

Sub-group	Effect	Units	Modified Blue	Modified Green	District Council Route	Do Minimum	Comments
Residential	Properties demolished	Number	13	3	2	1	Properties demolished on Blue route are Circa 1900. The cost of property acquisition and demolition is included in Group 6
	Noise	Number of houses experiencing increase of: more than 15 dB(A)L_{10} 10–15 dB 5–10 dB 3–5 dB	0 22 (5) 52 220	1 27 (6) 52 225	6 39 (10) 40 270	0 0 0 300	The changes in noise are the difference between the forecast for each option for 2008 and the existing levels. The units are dB(A)L_{10} 18 hr. 6 am–midnight. Allowance has been made for the presence of noise barriers in calculating these figures. The figures in parentheses are for those houses entitled to double-glazing
		Number of houses experiencing a decrease of: more than 15 dB(A)L_{10} 10–15 dB 5–10 dB 3–5 dB	0 138 266 300	12 248 142 350	11 181 117 200	0 0 0 0	
	Visual obstruction	Number of properties within 300m of centre line. Subject to:- Severe Significant Slight	3 8 15	4 6 21	3 10 18	No change No change No change	
	Visual intrusion		Slight intrusion to 20 houses	Moderate intrusion as River crossing is visible from 50 houses	As modified Green	No changes	Report of Landscape Advisory Committee gives more detailed information
	Severance a. Relief to existing severance b. Imposition of new severance		Substantial Slight	Moderate Severe	Moderate Slight	None None	
	Disruption during construction		10 houses within 100m of site will be affected for 6 months	10 houses within 100m of site will be affected for 6 months	12 houses within 100m of site will be affected for 6 months	None	
Industrial Premises			3 premises will experience an increase in noise of approx 4dB(A)L_{10} but since they are engaged in engineering work, this increased noise will not affect their operations	As modified blue	As modified Blue	No change	
Commercial Premises							
a. Office Buildings	Noise increase	Number subject to increase of more than 5dB(A)L_{10}	1 This building is double-glazed and the increase is, therefore, not noticeable to occupants	1	1	3	Average office occupancy on the new routes is 100–200 people. 'Do Minimum' route will affect 600–800 office workers
	Noise decrease	Number subject to decrease of more than 5dB(A)L_{10}	6	7	4	0	
	Visual obstruction	Number of properties within 300m of centre line Subject to: Severe Significant Slight	0 0 1	0 0 1	0 1 2	No change No change No change	
	Severance a. Relief to existing severance b. Imposition of new severance		Substantial improvement None	Moderate improvement None	Moderate improvement affecting 150 people	No improvement None	
	Disruption during construction		None	None	Access to 2 buildings affected for 3 months	Access to 3 buildings affected for one month	
b. Shops [1]	Noise increase	Number subject to increase of more than 5db(A)L_{10}	2	2	3	18	
	Noise decrease	Number subject to decrease of more than 5dB(A)L_{10}	27	31	29	8	
	Visual obstruction	Number of Properties within 300m of centre line Subject to: Severe Significant Slight	0 0 2	0 1 1	0 2 1	0 0 0	

NOTE:
[1] Entries in this sub-group relate to the interests of the people who own or work in shops. The effects on shoppers are given in Group 3.

Part A Section 2 Appendix 2

Group 2: Occupiers (continued)

Sub-group	Effect	Units	Modified Blue	Modified Green	District Council Route	Do Minimum	Comments
Shops (contd.)	Severance a. Relief to existing severance b. Imposition of new severance		Some improvement 1 shop severely affected	Some improvement Slight	Some improvement Slight	No improvement None	
	Disruption during construction		Pedestrian access to 20 shops affected for 3 weeks	Pedestrian access to 15 shops affected for 4 weeks	Pedestrian access to 20 shops affected for 4 weeks	Road Improvements will affect 50 shops for 3 months	
Schools and Hospitals							
a. Barchester Primary (233 pupils in 1978)	Noise Effect on 1 classroom where noise has been a particular problem	$dB(A)L_{10}$	Reduction of $5dB(A)L_{10}$	Reduction of $5dB(A)L_{10}$	Reduction of $2dB(A)L_{10}$	Increase of $3dB(A)L_{10}$	Based on maximum traffic flow expected on a normal working day within 15 years after opening
	Severance		Moderate improvement to existing access	As modified Blue	Slight improvement to existing access	Increased traffic flow will hinder access to school	
b. Horton Cottage Hospital (40 beds Accident Unit and Outpatients Dept. open weekdays only)	Noise increase	$dB(A)L_{10}$	$3dB(A)L_{10}$	$3dB(A)L_{10}$	$5dB(A)L_{10}$	No effect	Based on maximum flow as defined above
	Visual obstruction		Slight to Outpatients Dept.	Slight to Outpatients Dept.	Slight to Outpatients Dept.	No change	
	Disruption during construction		None	Access to Hospital disrupted for 1 month	Access to Hospital disrupted for 9 months	None	
Farming	Land take	**1** Number of Farms affected by land take	12	10	11	0	
		Hectares of land: Grade 2 Grade 3A	6·8 25·0	10·3 19·5	12·2 20·3	0 0	Based on **MAFF** land Classification. Compensation included in Group 6
Open Space							
a. Horton Golf Course (area 46 hectares)	Land take	Hectares	0	5·7	1·9	0	Effect on Users appears in Group 3
b. Low Road Methodist Chapel (area 2·8 hectares)	Land take	Hectares	0·6	0·6	1·0	0	Effect on Users appears in Group 3

NOTE:
1 In certain circumstances farms can be affected by severance where there is no land-take. In this case they should be included in this entry with a note to the effect in the comments column.

Page 129

Part A Section 2 Appendix 2

Group 3: Users of facilities

Sub-group: users of :—	Effect	Modified Blue	Modified Green	District Council Route	Do Minimum	Comments
a. Town Centre Shops High St./Market St. (100,000–160,000 shoppers per week)	Reduction of vehicle/pedestrian conflict	Reduces and diverts traffic sufficient to allow pedestrianisation	As modified Blue	As modified Blue	Existing vehicle/pedestrian conflict will increase with traffic growth in town centre	Based on updated County Council 1967 Shopping Study amended in County Structure Plan
b. Community Centre (i) Civic Theatre. (Used by average of 300 people each week in 1982)	Change in traffic noise in auditorium	5dB(A)L_{10} reduction	3dB(A)L_{10} reduction	3dB(A)L_{10} reduction	To maintain current noise level will require extensive sound proofing and air conditioning	Reductions are mainly in peak traffic periods and significant mainly at weekends
(ii) Public Library. (Used by average of 1,200 people each week in 1982)	Change in traffic noise in reading room	3dB(A)L_{10} reduction	3dB(A)L_{10} reduction	3dB(A)L_{10} reduction	Existing noise will increase with traffic growth	
(iii) Day Care Centre. (Used by average of 600 old age pensioners and helpers each week in 1982)	Effect on access for the elderly	35–40% reduction in traffic	35–40% reduction in traffic	35–40% reduction in traffic	40% increase in traffic will make pedestrian access more difficult	Average age of members is 74 years
c. Warren Street shops (60,000 shoppers per week)	Convenience of customers	No facilities on new route	As modified Blue	As modified Blue	No effect	
d. Horton Golf Club (382 members in 1981)	Reduction of amenity due to land take	No effect	Reduced to 17 holes. Substantial redesign and construction could restore it to 18 holes but would require closure for 2 growing seasons	Remains at 18 holes but edge of course adjacent to 12th hole is taken	No effect	No other golf courses are locally available
e. Sailing Club (106 members in 1981)	Reduction in amenity (visual intrusion, sailing conditions, etc.)	7·6m embankment and river bridge effectively prevents sailing on last 200m of course	8·5m embankment and river bridge effectively prevents sailing on last 100m of course	7m embankment and river bridge cut sailing course approx. in half	No effect	Few sailing clubs in the area. Recently built club house supported by Sports Council
f. Horton Hunt (236 members in 1980)	Severance	2 fox runs north of town severed	As modified Blue	As modified Blue	No effect	
g. North Waxton Ornithological Society (57 members in 1981)	Loss of abandoned gravel pits	Gravel pits partly filled. Proximity of new road will disturb birds	As modified Blue	Eastern part of gravel pits filled. Proximity of new road will disturb birds	No effect	
h. Barchester Fishing Club (85 members in 1981)	Loss of fishing rights in gravel pits	Gravel pits partly filled preventing fishing	As modified Blue	Eastern part of gravel pits filled leaving only a quarter of original area for fishing	No effect	
i. Low Road Methodist Chapel (Average congregation 35)	a. Noise increase	5dB(A)L_{10} increase	3dB(A)L_{10} increase	9dB(A)L_{10} increase	No effect	These increases are less on Sundays
	b. Visual obstruction	6m embankment 30m from church	As modified Blue	8m embankment 25m from church	No effect	
	c. Severance from main part of town	Slight severance	Slight severance	Moderate severance	No effect	Land take effects appear in Group 2. Compensation in Group 6

Page 130

Group 4: Policies for conserving and enhancing the area
(Views expressed are those of the relevant Authority unless otherwise stated)

Policy	Authority	Interest	Modified Blue	Modified Green	District Council Route	Do Minimum	Comments
a) To protect the Hill Street Outstanding Conservation Area	Dartshire CC / Barchester DC	Improvement of the environmental quality of the conservation area and reduction of pedestrian/vehicle conflict	Reduces and diverts traffic sufficient to allow pedestrianisation	As modified Blue	As modified Blue	Traffic levels will increase with time to detriment of cobble square	DOE designated area as Outstanding in 1976. Contains one Grade I and three Grade II listed buildings
b) To protect other listed buildings outside Conservation Area	DOE / Dartshire CC / Barchester DC	Effect on Wattle Hall a grade II listed building	Road in 1m cutting 500m from house	Road on 1.3m embankment 300m from house	No effect	No effect	Listing is based on interior fittings and ceilings
c) To preserve Antiquities	DOE / Dartshire CC / Barchester DC	Effect on tumuli: number destroyed	3	3	2	0	The area has numerous tumuli of the same period. There will be opportunity for the Dartshire Archeological Society to excavate
d) To protect Landscape in Avon Valley	Dartshire CC / Barchester DC / Orford PC / National Tourist Board	Effect on view from Orford Church referred to in Wilton's Poem "Across the Lea"	No effect	Road on 1m embankment 600m from Church (no comment received from County Council)	Road on 2m embankment 500m from Church	No effect	Report of Landscape Advisory Committee covers Orford Church which has a Saxon Arch and Georgian Choir Stalls and is linked in legend to Hereward the Wake
e) To restore derelict land in the Avon Valley	DOE / Dartshire CC / Barchester DC	Restoration of abandoned gravel pits. Hectares affected	6 unaffected, 9 can be used for spoil tips and restored	8 unaffected, 7 can be used for spoil tips and restored	10 unaffected, 5 can be used for spoil tips and restored	No effect	See also Dartshire CC Policy on Country Park. See also British Waterways Board Policy on canal network
f) To create a Country Park Leisure Centre adjacent to River Avon West of Barchester	Proposed and supported by:- Dartshire CC / Barchester DC / Sports Council / Countryside Commission Opposed by:- Horton PC	To create a Country Park and Leisure Centre along river bank and to incorporate disused gravel pits	Would prevent the creation of Country Park. Water based sports could not be developed	As modified Blue	Area of possible Park would be much reduced and overshadowed by road on high embankment	No effect	The Country Park appears as a policy in the County Structure Plan and District Council Local Plan. The creation of a Leisure Centre has potential for grant aid
g) To maintain and improve national canal network	British Waterways Board	Use of disused gravel pits as regulatory reservoirs	Less potential capacity for use as balancing reservoir	As modified Blue	Substantially less potential capacity for use as balancing reservoir	No effect	
h) To protect the habitat of rare plants	Dartshire Botanical Society [1]	Habitat of Cypripectium leitchum (orchid)	Destroys habitat	No effect	No effect	No effect	Only 4 known habitats in England. A full ecological report is available.

Note:
[1] Assessment of particular impacts by local bodies should be included in the framework only where they have expert knowledge which is not available elsewhere.

Part A Section 2 Appendix 2

Group 5: Transport, development and economic policies
(Views expressed are those of the relevant Authority unless otherwise stated.)

Policy	Authority	Interest	Modified Blue	Modified Green	District Council	Do Minimum	Comments
Transport							
a) To improve trunk roads to ports	Department of Transport	Ease of Access from manufacturing Centre to the port	Big Improvement	Big Improvement	Some Improvement	Increasing delays expected	White Paper on Road Policy 1980
b) To relieve local traffic problems in Barchester	Dartshire CC	Convenience of local traffic	Most effective; removal of through traffic will give scope for local traffic management measures	As modified Blue	Slightly less effective. Off peak traffic may continue to use existing route	No benefit	Dartshire CC is Highway Authority
c) To concentrate heavy goods vehicles on suitable roads	Department of Transport	% Transfer of HGVs to new route from existing route	40 – 60%	35 – 55%	20 – 30%; junction layout and location discourages transfer	No effect	
d) To improve safety and to upgrade the London to Camelot line	British Rail	Removal of Heton level crossing	Removes the need for the crossing	Crossing remains	As modified Green	As modified Green	Removal of crossing would obviate the need for local authority small scheme improvement at Heton scheduled for 1988 cost £500,000 at 1979 prices
e) To maintain river Avon navigation	British Waterways Board	Temporary effect of bridge construction on navigation	Slight reduction of head room will occur at one bridge construction site for short periods	Slight reduction of headroom will occur at bridge construction sites for short periods	As modified Blue	None	Licence under Navigation and Waterways Act required
f) To maintain viable rural bus transport system in south Dartshire	Dartshire CC Bus Operators	Effect on service reliability	Improvement	Improvement	Improvement	Increased traffic delays will reduct the bus-service reliability	
Development & Economic							
a) To develop Barchester as Regional Shopping Centre	Dartshire CC Barchester DC	Improve accessibility to and the amenities of shopping centre	Improves access and provides for pedestrianisation in area, but will disbenefit shops in Warren Street accelerating the decline of this twilight area	As modified Blue	As modified Blue but the effect on Warren Street shops is less severe	Current traffic congestion and delivery difficulties will increase	Policy contained in County Structure Plan and District Council Local Plan
b) To limit growth in south of County and encourage new employment and housing in villages of Scapton, Haydon, and Wettering	Dartshire CC	Effect on rural northern sector of Dartshire	Improves access to Scapton, Haydon and Wettering as well as north of County	No effect	Improves access to Scapton, Haydon and Wettering as well as north of County	No effect	County Structure Plan policy
c) To safeguard identified commercially workable gravel resources in the River Avon Valley	DOE	Gravel beds underlying river flood plan to west of Barchester	None	3·2 hectares affected	2·8 hectares affected	None	County Structure Plan policy. Time would permit the extraction of the gravel prior to construction
d) To encourage all existing non conforming industry to relocate and all new industry to locate on the Barchester Industrial Estate	Proposed by Barchester DC Opposed by Dartshire CC	Effect on access to Industrial Estate	Improves access	Improves access	No effect	No effect	Both the District and County Councils favour concentration of new and non conforming industry on Industrial estates, but Dartshire would prefer growth at Blaydon City rather than Barchester. Non conforming industry is that not compatible with the general land use in the area

Page 132

Part A Section 2 Appendix 2

Group 6: Financial Effects

Sub-group	Interest	Units	Modified Blue		Modified Green		District Council		Do Minimum	Comments
Department of Transport [1]	Construction costs	£m (PVC)	4·70		4·60		5·80		0·70	Costs are discounted from years of expected expenditure to 1979 at 1979 prices (PVC = present value of costs, PVB = present value of benefits, NPV = net present value)
	Land costs	£m (PVC)	1·90		2·10		2·00		0·40	
	Compensation costs	£m (PVC)	0·30		0·30		0·40		0·08	
	Maintenance costs	£m (PVC)	0·03		0·03		0·03		0·01	Excess maintenance cost due to additional length of road or improved lighting, signing etc
	Total cost	£m (PVC)	6·93		7·03		8·23		1·19	
Total quantified monetary benefits		£m (PVB)	High 7·27	Low 3·80	High 6·18	Low 3·62	High 5·95	Low 3·32	0	Includes savings in time, vehicle operating costs and accidents. Taken from Group 1
Net present value compared to do minimum		£m (NPV)	+1·53	−1·94	+0·34	−2·22	−1·09	−3·72	0	

1 Construction costs should include preparation and supervision costs.

ANNEX 6

RULES FOR THE CALCULATION OF COLOUR CODING

1   The system was developed to provide a simple basis on which the Treasury could delegate to the Department responsibility for approving some schemes which show a negative net present value, but which have clear environmental net benefits. The current delegation applies to schemes which have a colour code of Green or Green/Yellow and a negative net present value not exceeding £0.5m. We are informed that there has been a small number of schemes falling within this category over the last 10 years.

2   Colour coding is done at two stages; first, before a decision is taken to present it as an option for Public Consultation, and again before it is presented as a Preferred Scheme at the Public Inquiry stage.

3   The rules are formulated in such a way that they give greater weight to demolition, noise and visual obstruction than to the other effects. There is also no attempt to avoid double-counting. Nevertheless this approach was adopted on the basis of analyses of the environmental effects which have been seen as important influences in past decisions. Greater weight is given to disbenefits than to benefits, and the score necessary to achieve a "Green" coding is greater than that which attracts "Red". Taken together, these controls will tend to understate any given scheme's environmental quality.

4   The calculation of colour codes takes account of the following environmental effects:

   i)   demolition of houses

   ii) noise changes greater than 3dB(A) affecting

        a)   houses
        b)   shops

   iii) moderate or high visual obstruction affecting

        a)   houses
        b)   shops

   iv) moderate or high visual intrusion

   v)   moderate or severe severance or substantial relief to existing severance for an appreciable number of people.

   vi) substantial effect on policies to conserve or enhance the environment.

   Items ii), v), and vi) may include environmental gain as well as damage.

5   The detailed procedure for the calculation of colour codes is as follows:

PROCEDURE FOR ENVIRONMENTAL COLOUR CODING

Direct Benefits/Disbenefits

(i) Add together the numbers of houses and shops relieved of noise by more than 3dB(A)[1].

(ii) Add together the number of houses and shops exposed to extra noise over 3dB(A) or visual obstruction[2] and increase this number by 25%. Add on 1½ times the number of houses demolished.

(iii) Subtract 2 from 1 to give preliminary colour coding as follows:-

over + 100	green
+25 to + 100	green/yellow
+25 to -25	yellow
-25 to -50	yellow/red
over -50	red

Indirect Benefits/Disbenefits

(iv) Compare the effects on environmental policies. If the net effect is substantially positive or negative change the colour code by one step.

(v) Consider the scale of visual intrusion. If this assessed as "high", change the colour code by one step towards red.

(vi) Consider the scale of community severance relieved or imposed. If this is classed as substantial relief of severance or severe new severance for an appreciable proportion of the inhabitants, change the colour code by one step.

Minor Disbenefits

(vii) If disbenefits assessed as moderate are expected against all headings of policies, community severance and visual intrusion change the colour code by one step towards red.

6   The application of the rules involves a mixture of technical measurement and subjective appraisal. The assessment of visual intrusion, severance and effects on policies, for example may leave considerable scope for judgement. The inclusion of item vi - effect on policies - in the table means that double-counting is almost inevitable, with any of the preceding items, each of which may be the subject of local or national policy. (The same problem, incidentally, arises under the MEA

---

[1] Double the number if noise changes are 15dB(A) or more.
[2] Slight visual obstruction should be ignored.

Framework itself, for the same reason). However, the Department considers the level of potential double-counting to be manageable, since (in general) it results in a slightly greater emphasis being attached to disbenefits than to benefits and the uses to which colour-coding is put are purely administrative, to do with internal allocation of responsibility between the Treasury and the Department itself.

GLOSSARY

GLOSSARY

BLIGHT:

The reduction in value of a property and increased difficulty in selling it which can arise from uncertainty over development proposals.

COBA:

Standing for Cost-Benefit Analysis, the name given to the computer program used by the Department of Transport to produce economic evaluations of its road schemes.

COLOUR CODING:

A system used within the Department of Transport to obtain an overall assessment of environmental advantages and disadvantages of trunk road schemes, used for the purpose only of determining various delegation powers.

COMMUNITY SEVERANCE:

The extent to which a road or proposed improvement cuts across established patterns of community activities, causes extra delays and diversions to local trips in consequence, and may impede development.

CONSTRAINT:

Used in this report in the context of policies which impose certain limits to development or change. For example a requirement that certain environmental features or heritage sites cannot be destroyed or devalued, or alternatively such impacts must be mitigated or reinstated elsewhere.

CORRIDOR:

A part of a road network which displays a clear and dominant linear pattern of travel demands. For example a motorway connecting two built-up areas, the old adjacent "A" road, and, perhaps, a railway link serving broadly the same destinations.

DISCOUNT RATE:

The annual percentage rate which reduces the values of benefits and costs occurring in future years to equivalent present day values.

DISCOUNTING:

The process by which economic benefits and costs which are expected to occur in the future are reduced to reflect the fact that they have a lower value relative to the same benefits and costs which arise today.

DO-MINIMUM:

The existing transport network as amended to take account of planned and expected improvements against which the full scheme option is appraised.

DO-NOTHING:

The existing network with no changes. Occasionally this will be a more appropriate base than the "Do-minimum" against which to compare the scheme.

ENVIRONMENTAL STATEMENT:

A document describing the environmental characteristics of a site and surrounding location of a proposed development, the scheme itself and predicted environmental effects, proposed mitigation, and the methodology employed in the assessment and the resulting data. Under the European Community Directive 85/337 developers are required to produce a formal Environmental Statement for certain types of development when planning consent is applied for. For trunk road investments this Directive was given legal effect by amendments to the 1980 Highways Act.

FIXED TRIP MATRIX:

A matrix of origin-destination (O-D) trips which remains unchanged as a result of a transport improvement. A variable trip matrix is one in which the demand for O-D trips varies as a result of a transport improvement.

FRAMEWORK:

A structured statement of the effects of a scheme. It is usually presented in the form of a table with different options at the head of the columns and the various effects and groups of people affected in each row. (When, in this report, the term is written as "the Framework", ie with a capital F, it is a reference to the formal presentation tool developed after the first SACTRA report and still used by the Department, primarily at Public Consultation and Public Inquiry stages.)

HYPOTHETICAL VALUES:

Values derived from people's statements of their preferences. The impacts that can be addressed in hypothetical valuation are those perceived at an individual level rather than a community level.

LOCAL MODEL VALIDATION REPORT:

A report explaining the extent and format of a traffic model, the sources of its traffic data, the assumptions and relationships employed, and - very importantly - confirming its predictive capabilities by validating against additional independent traffic data.

LOCAL PLAN:

These are normally prepared and adopted by District planning authorities, although sometimes by County planning authorities. Local Plans develop the policies and general proposals of Structure Plans and relate them to precise areas of land.

MANUAL OF ENVIRONMENTAL APPRAISAL:

Manual of Environmental Appraisal (MEA). The name given to the manual which the Department of Transport currently uses to assess environmental impacts of schemes.

MITIGATION:

Action taken to reduce any adverse environmental impacts of a road scheme.

NET PRESENT VALUE:

The difference between the accruals of costs and benefits of a scheme, both of which are on the same price basis and have been discounted to a common year. Often shortened to NPV.

NATIONAL ROAD TRAFFIC FORECAST:

National Road Traffic Forecasts (NRTF), Great Britain (1989) are the Department of Transport's forecasts of growth in vehicle kilometres of traffic nationally from 1988 to 2025. They are not meant to apply in themselves to any particular road link, but, rather to serve as a control total for the forecasts produced for the appraisal of particular road improvement schemes.

OPTION:

A proposed course of action designed to meet specific objectives and usually compared with other courses of action which are the subject of assessment.

PREFERRED ROUTE:

The route selected from several options by the Secretary of State for Transport for further detailed design and appraisal, after consideration of the Technical Appraisal Report and the Report on the Public Consultation.

PUBLIC CONSULTATION:

>The stage at which the Department of Transport presents material to the public to inform that a road scheme is under consideration, to indicate possible options and likely consequences, and to call for views.

PUBLIC INQUIRY:

>The opportunity, laid down by Statute, for any objectors to the published orders of a scheme to put their case in front of an independent Inspector, who may recommend changes in the light of such objections and the Department's own case.

REVEALED VALUES:

>Values deduced from observation of actual human behaviour.

ROAD PROGRAMME:

>The list of trunk road schemes (new roads or improvements to existing roads) which the Department intends to construct in so far as appraisal shows them to be worthwhile.

SCHEME:

>An improvement proposal for a road or public transport.

SHADOW COSTS:

>Costs, used in cost-benefit analysis, which are estimated more closely to correspond to real long term resource costs to the economy than would the market price. The term may also be used to refer to the costs which are incurred due to adhering to some binding policy constraint, for example in mitigating environmental impact.

STRUCTURE PLAN:

>The broad policies for the development and use of land at the county level (including measures for the improvement of the physical environment and traffic management) to be adopted by local authorities.

SUSTAINABILITY:

>Humanity's ability to ensure that it meets the needs of the present without compromising the ability of future generations to meet their own needs.

TECHNICAL APPRAISAL REPORT:

The report on the technical aspects of the highway problem and the viable alternative solutions. It evaluates these options on engineering, costs, economic, and environmental grounds and recommends the alternatives to be put to public consultation.

TRAFFIC APPRAISAL MANUAL:

Traffic Appraisal Manual (TAM). A manual produced by the Department of Transport providing advice, and describing recommended and mandatory practices on all aspects of traffic appraisal, including obtaining and handling data, model building and producing local forecasts.

TRAFFIC FLOW:

The number of vehicles passing a particular point on a road within a given time period.

TRIP MATRIX:

The numbers of journeys demanded between different origins and destinations within a given time period, presented in the form of a grid with origins along the side and destinations across the top.

TRUNK ROAD:

Part of the national network of all-purpose through routes for which the Secretary of State is responsible.

WEIGHTING:

A method of comparing options by ascribing relative scores to the different attributes. These weights should reflect the options' predicted impacts, taking account both of the relative performance of each option under different objectives or impacts, and the relative importance of the various impacts.